T0348704

Learn Data Science Using Python

A Quick-Start Guide

Engy Fouda

Apress®

Learn Data Science Using Python: A Quick-Start Guide

Engy Fouda
Hopewell Junction, NY, USA

ISBN-13 (pbk): 979-8-8688-0934-7   ISBN-13 (electronic): 979-8-8688-0935-4
https://doi.org/10.1007/979-8-8688-0935-4

Managing Director, Apress Media LLC: Welmoed Spahr
Acquisitions Editor: Celestin Suresh John
Development Editor: James Markham
Coordinating Editor: Gryffin Winkler

Cover designed by eStudioCalamar

Cover image by GarryKillian on Freepik.com

Distributed to the book trade worldwide by Apress Media, LLC, 1 New York Plaza, New York, NY 10004, U.S.A. Phone 1-800-SPRINGER, fax (201) 348-4505, e-mail orders-ny@springer-sbm.com, or visit www.springeronline.com. Apress Media, LLC is a California LLC and the sole member (owner) is Springer Science + Business Media Finance Inc (SSBM Finance Inc). SSBM Finance Inc is a **Delaware** corporation.

For information on translations, please e-mail booktranslations@springernature.com; for reprint, paperback, or audio rights, please e-mail bookpermissions@springernature.com.

Apress titles may be purchased in bulk for academic, corporate, or promotional use. eBook versions and licenses are also available for most titles. For more information, reference our Print and eBook Bulk Sales web page at http://www.apress.com/bulk-sales.

Any source code or other supplementary material referenced by the author in this book is available to readers on GitHub (https://github.com/Apress/learn-data-science-using-python) and at www.apress.com/9798868809347. For more detailed information, please visit https://www.apress.com/gp/services/source-code.

If disposing of this product, please recycle the paper

To my sweet genius Areej, keep smiling and continue shining like the star you are.

To Hesham, who never gives up on me despite I sometimes drive you crazy.

To my parents, I hope you are now proud of me.

To my brothers, family, and friends, you always lend an ear to my venting.

Table of Contents

About the Author

Engy Fouda is an adjunct lecturer at SUNY New Paltz University, teaching Introduction to Data Science using SAS Studio and Introduction to Machine Learning using Python. She is an Apress and Packt Publishing author. Currently, she teaches SAS fundamentals, intermediate SAS, advanced SAS, SAS SQL, introduction to Python, Python for data science, Docker fundamentals, Docker enterprise for developers, Docker enterprise for operations, Kubernetes, and DCA and SAS exams test-prep course tracks at several venues as a freelance instructor.

She also works as a freelance writer for Geek Culture, Towards Data Science, and Medium Partner Program. She holds two master's degrees: one in journalism from Harvard University, the Extension School, and the other in computer engineering from Cairo University. Moreover, she earned a Data Science Graduate Professional Certificate from Harvard University, the Extension School. She volunteers as the chair of Egypt Scholars board and is former executive manager and former Momken team leader (Engineering for the Blind). She is the author of the books *Learn Data Science Using SAS Studio* and *A Complete Guide to Docker for Operations and Development* published by Apress and a co-author of *The Docker Workshop* published by Packt.

About the Technical Reviewer

 **Suvoraj Biswas** is a subject matter expert and thought leader in enterprise generative AI, specializing in its architecture and governance. He authored the award-winning book *Enterprise Generative AI Well-Architected Framework & Patterns*, which has received global acclaim from industry veterans for its original contribution in defining a well-architected framework for designing and building secure, scalable, and trustworthy enterprise generative AI across industries.

Professionally, Suvoraj is an architect at Ameriprise Financial, a 130-year-old Fortune 500 organization. With over 19+ years of experience in enterprise IT across India, the United States, and Canada, he has held architectural and consulting roles at companies like Thomson Reuters and IBM.

His expertise spans secure and scalable enterprise system design, multicloud adoption, digital transformation, and advanced technology implementation, including SaaS platform engineering, AWS cloud migration, generative AI adoption, and smart IoT platform development.

His LinkedIn profile is https://www.linkedin.com/in/suvoraj/.

Introduction

This book shows you in step-by-step sequences how to use Python to accomplish data cleaning, statistics, and visualization tasks even if you do not have any programming or statistics background.

The book's case study predicts the presidential elections in the state of Maine, which is a project I did at Harvard University. Chapter 1 explains the case study in more detail. In addition to the presidential elections, the book provides real-life examples, including analyzing stocks, oil and gold prices, crime, marketing, and healthcare. You will see data science in action and how to accomplish complicated tasks and visualizations in Python.

You will learn, step-by-step, how to do visualizations. The book includes explanations of the code with screenshots of every step. The whole book follows the paradigm of data science in action to demonstrate how to perform complicated tasks and visualizations in Python in an easy to follow way and with real big data for interesting hands-on-labs.

It will provide readers the required expertise in data science and analytics using Python:

- Import and export raw data files

- Manipulate and transform data

- Combine datasets

- Create summary statistics and reports

- Identify and correct data, syntax, and programming logic errors

- T-test

- Linear regression

- Visualizations

Who Should Read This Book?

The primary audience of this book is students who are newbies to data science and might not have deep programming experience. Consequently, university professors can use it as a handout for their courses. Also, technical instructors, like myself, who teach professionals from various industry sectors at certified training centers, can teach from it and use it as the courses' textbook.

Also, system analysts and scientists who are experienced but new to Python will find faster and more efficient tools to achieve their daily tasks. Moreover, data journalists and investigative reporters will find the book easy to follow and can generate essential visualizations fast.

How Is This Book Organized?

The book has three parts. The first part, Chapters 1, 2, and 3, gets you familiar with the SAS interface and the basic essential tasks. Chapter 1 focuses on drawing a general idea of the case study of the presidential elections project in the state of Maine and its outputs. Throughout the book, you will learn how to get those output charts and analytics in a step-by-step manner.

Then the second part, Chapters 4, 5, and 6, focuses on more advanced programming aspects. The third part, Chapter 7, takes you from analyzing historical data to predicting the future.

All the example code is in the "Example Code" folder, and datasets required for this book are in the "Datasets" folder.

CHAPTER 1

Data Science in Action

This chapter presents a case study examining voter data analysis in the state of Maine. Stemming from a project undertaken during my master's program at Harvard University in the Fall of 2016, under the guidance of Professor Larry Adams, the project aimed to forecast presidential election results nationwide. My task was to predict Maine's 2016 and 2020 election outcomes.

The project unfolded in two phases, initially focusing on predicting the 2016 election results and validating the findings against actual outcomes before proceeding to the second phase. The second phase incorporated new data generated in 2016 to forecast the 2020 results: hence, some sections of this book feature 2016 data. Where feasible, I gathered historical data relevant to the analysis, including election data dating back to 1960.

Voter demographics were delineated by age, gender, education, demographics, and race. Issue categories pertinent to the state, including the economy, education, environment, healthcare, and gun control, were identified through comprehensive research. Moreover, I examined county ballot topics to gauge their potential impact on the presidential election.

Various prediction methodologies and algorithms, including Monte Carlo and Bayes, as well as statistical tests like T-test and chi-square, were explored, leveraging party voting patterns since 1960, poll maccuracy, and electoral votes. In the past, FiveThirtyEight and other forecasting platforms were compared, and it was found that the prediction made in 2016 was accurate. This project provided valuable insights into translating cognitive

© Engy Fouda 2024
E. Fouda, *Learn Data Science Using Python*, https://doi.org/10.1007/979-8-8688-0935-4_1

features into numerical data for analysis, culminating in informed decision-making based on trend measurement and pattern analysis—a skillset honed through subsequent data science endeavors.

For this project, I used both SAS and R programming languages. In 2020, I published a book, *Learn Data Science Using SAS Studio*, with Apress. The book has been a hit and phenomenally successful; therefore, Apress and I decided to rewrite it using Python.

Data Science Process

The data science process initiates by formulating a question or hypothesis; then by collecting pertinent raw data; followed by data cleaning and exploration; modeling and evaluation; and finally, the deployment, visualization, and communication of findings, as illustrated in Figure 1-1.

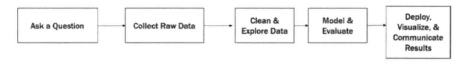

Figure 1-1. *Data science process*

Questions in various fields differ, for example, in politics, where inquiries revolve around predicting electoral outcomes, like whether Trump will win in Maine in 2016 and 2020, or in the context of Facebook, for exploration of strategies to increase user engagement.

The second stage involves gathering raw data, manifested by acquiring comprehensive voter information, including age, race, education, income, gender, industry, historical ballot data, and voting results. The accuracy of the predictions increases when there is more historical data. Furthermore, gathering information about population distribution over the years is crucial.

Subsequently, the third stage entails cleaning the raw data; addressing issues like managing missing values, outliers, and repeated rows; and correcting misspellings while standardizing data types and formats across columns.

The fourth stage involves experimenting with multiple models and comparing their tailored performance to the given problem. For instance, I utilized Monte Carlo and Bayes algorithms in the presidential election scenario.

The final stage involves visualizing and articulating the results in accessible language within reports. This step is paramount as it culminates in addressing the initial question that instigated the entire process, thereby fulfilling the primary objective of the endeavor.

Case Study: Presidential Elections in Maine

As noted earlier, the data science journey commences with a query. In this project's scope, my inquiry focused on determining whether Donald Trump will win the state of Maine during the presidential elections of 2016 and 2020 or not.

Population

The next step involves gathering relevant data extensively. First, I focused on understanding the population distribution across Maine's counties, utilizing information from the US Census Bureau. This exploration revealed that population distribution is unique, with vast areas either sparsely populated or inhabited by solitary individuals. While the tiny dots in the southern region may seem insignificant, each has over 5,000 people. Thus, be cautious against misinterpretations from maps disseminated by presidential campaigns or mainstream media outlets.

Subsequently, the logical progression was to acquire voters' information. While some states offer free access to their voters' databases, Maine operates differently, selling such databases exclusively to political parties. Consequently, I contacted the Secretary of State's office for assistance.

In response, I was informed that access to Maine's Central Voter Registration system required affiliation with one of five specific categories: a candidate or campaign representative, a party affiliate, a participant in a referendum campaign, an organizer of get-out-the-vote efforts, or an elected official for authorized purposes. Costs were determined based on the number of records requested, as outlined in Title 21-A, Section 196-A. Notably, a statewide voter file comprising one million records incurred a fee of $2,200.

Following several email exchanges elucidating the research nature of my request and providing requisite verifications, the office graciously offered a complimentary DVD containing the necessary information, with partial data relevant to my research, and the voters' personal data, such as last names; addresses were removed from the datasets.

The DVD's primary table features voters' details, as depicted in Figure 1-2, with columns including first name, year of birth, enrollment code, special designations, registration date, congressional district, county ID, modification date, and last statewide election with VPH.

	A	B	C	D	E	F	G	H	I
1	FIRST NAME		YOB	ENROLL	DESIGNATE	DT ACCEPT	CG	CTY	DT CHG	DT LAST VPH
2			1913	R		12/5/1935	2	01AND	11/26/2008	
3			1918	R		9/8/1947	2	01AND	5/22/2008	6/10/2008
4			1925	D		10/14/1952	2	01AND	5/17/2010	11/2/2010
5			1928	D		11/8/2005	2	01AND	4/25/2012	11/4/2008
6			1929	R		10/20/2009	2	01AND	6/13/2012	6/14/2016
7			1932	FIELD NAME			2	01AND	12/31/2005	11/6/2012
8			1935	FIRST NAME			2	01AND	8/18/2010	11/4/2014
9			1944	YEAR OF BIRTH			2	01AND	2/8/2016	11/4/2014
10			1949	ENROLLMENT CODE			2	01AND	11/7/2007	6/14/2016
11			1950	SPECIAL DESIGNATIONS			2	01AND	3/29/2016	11/6/2012
12			1963	DATE ACCEPTED (DATE OF REGISTRATION)			2	01AND	12/30/2015	11/4/2008
13			1986	CONGRESSIONAL DISTRICT			2	01AND	7/25/2012	
14			1995	COUNTY ID			2	01AND	12/18/2015	
15			1949	DATE CHANGED			2	01AND	10/24/2012	
16			1980	DATE OF LAST STATEWIDE ELECTION WITH VPH			2	01AND	8/12/2015	
17			1964				2	01AND	12/31/2005	11/4/2014
18			1938	R		9/28/1964	2	01AND	12/31/2005	6/14/2016
19			1950	U		10/16/2006	2	01AND	12/31/2005	11/4/2014
20			1955	D		8/10/1998	2	01AND	9/8/2016	11/4/2014
21			1960	U		01/01/1850	2	01AND	12/7/2009	11/4/2008
22			1966	U		2/11/2000	2	01AND	5/15/2012	11/4/2014

Figure 1-2. *Voters' information*

The subsequent table summarizes registered and enrolled voters, illustrated in Figure 1-3. Its columns include the county name, municipality name, ward precinct, congressional district, state senate district, county commissioner district, party affiliation, and total count. Party affiliations documented in the file encompass Democratic, Green Independent, Libertarian, Republican, and unenrolled.

	A	B	C	D	E	F	G	H	I	J	K	L	M
	COUNTY	MUNICIPALITY	W/P	CG	SS	SR	CC	D	G	L	R	U	TOTAL
2	AND	AUBURN	1-1	2	20	62	5	625	121	28	355	622	1751
3	AND	AUBURN	1-1	2	20	64	5	115	20	5	139	143	422
4	AND	AUBURN	1-1	2	20	64	6	321	34	7	317	342	1021
5	AND	AUBURN	2-1							28	106	250	626
6	AND	AUBURN	2-1							22	383	630	1691
7	AND	AUBURN	2-1							9	287	383	1002
8	AND	AUBURN	3-1							34	100	289	676
9	AND	AUBURN	3-1							12	262	327	932
10	AND	AUBURN	3-1							9	348	543	1341
11	AND	AUBURN	3-1							1	84	138	336
12	AND	AUBURN	4-1							24	118	313	792
13	AND	AUBURN	4-1							3	147	265	665
14	AND	AUBURN	4-1							13	409	507	1364
15	AND	AUBURN	5-1							33	169	447	1115
16	AND	AUBURN	5-1							19	309	501	1339
17	AND	AUBURN	5-1							4	180	310	680
18	AND	DURHAM	1-1							15	967	1297	3297
19	AND	GREENE	1-1							28	903	1271	3257
20	AND	LEEDS	1-1							9	468	705	1743
21	AND	LEWISTON	1-1							29	353	501	1699
22	AND	LEWISTON	1-1	2	21	00	2	1200	122	63	275	909	2629
23	AND	LEWISTON	2-1	2	21	59	2	1538	163	45	849	1169	3764
24	AND	LEWISTON	3-1	2	21	59	2	556	36	3	88	287	970
25	AND	LEWISTON	3-1	2	21	60	1	1429	233	202	359	1324	3547

Overlaid field-name box:

FIELD NAME
COUNTY NAME
MUNICIPALITY NAME
WARD PRECINCT
CONGRESSIONAL DISTRICT
STATE SENATE
STATE REPRESENTATIVE
COUNTY COMMISSIONER DISTRICT
DEMOCRATIC
GREEN INDEPENDENT
LIBERTARIAN
REPUBLICAN
UNENROLLED
TOTAL

Figure 1-3. *Registered and enrolled voters table*

The initial dataset was full of erroneous values and outliers. For instance, one voter's age was listed as 220 years despite their birthdate indicating an age of approximately 67 years—additionally, some voter information was entirely blank. As emphasized earlier, it is imperative to meticulously clean your data by addressing outliers and missing values, formatting inconsistencies, and conducting thorough data exploration. In Chapter 6, we will perform the data cleaning together in step-by-step labs.

Furthermore, I thoroughly searched and gathered data from various sources to ensure I had enough historical data. I also obtained additional tables from the US Census Bureau website:

`https://www.census.gov/data/tables/time-series/demo/voting-and-registration/p20-585.html.`

Subsequently, I categorized voters based on gender, age, and race for further analysis.

Gender

As depicted in Figure 1-4, my analysis revealed that the number of registered female voters surpasses that of registered male voters in Maine. However, the percentage disparity between the two genders is between ±2%. Regarding the strategic implications for political campaigns, representatives could consider leveraging this information by, for example, wearing the cancer awareness pink ribbon to resonate with female voters, given their historically higher turnout than men. This recommendation highlights the significant influence that data and statistics wield, shaping speech topics and influencing political representatives' attire.

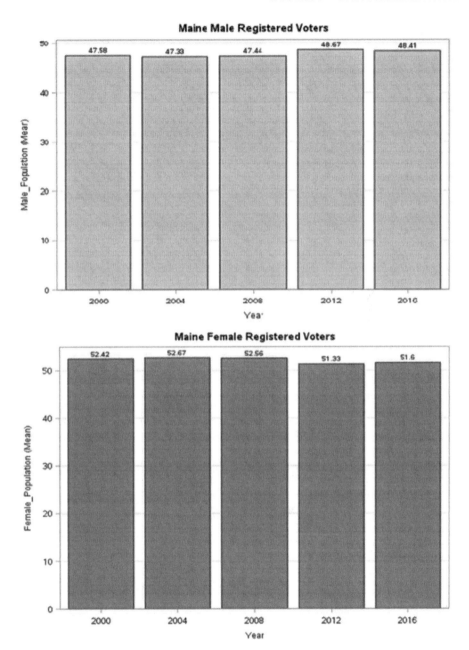

Figure 1-4. *Maine registered voters grouped by gender*

Race

Concerning racial demographics in Maine, over 96% of the population identifies as white, as indicated in Figure 1-5. Consequently, I categorized Black, Asian, and Hispanic voters as non-white registered voters.

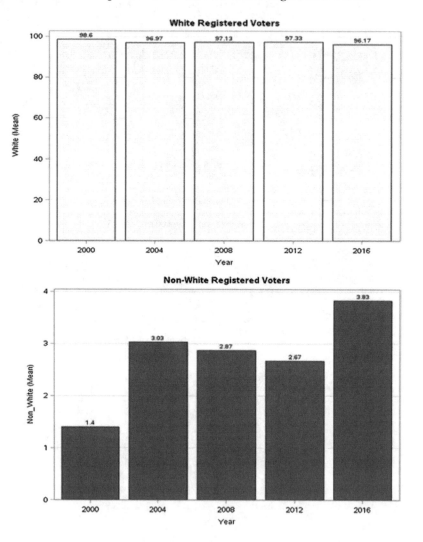

Figure 1-5. *Maine registered voters grouped by race*

Age

In 2016, there was a notable decline in registered voters over 65, as depicted in Figure 1-6 (note the varying scales). Conversely, there was an uptick in the 18–24 and 25–44 age brackets. This observation suggests a need for adjustments in speech topics to resonate with a broader voter base. For instance, campaign representatives should prioritize issues like student loans and home mortgages instead of prioritizing medical insurance and retirement funding discussions.

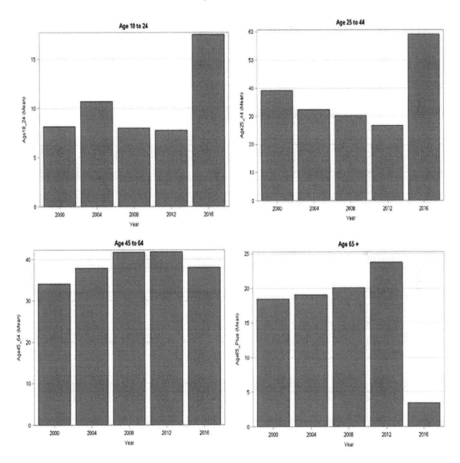

Figure 1-6. *Maine registered voters grouped by age*

Voter Turnout

I examined historical voter turnout to gauge the number of individuals who braved Maine's snowy streets to cast their votes. Figure 1-7 illustrates voter turnout percentages from 2000 to 2016, with the Democratic Party emerging as the victor.

If you try this exercise with a swing party, the columns' colors will change to reflect the winning party.

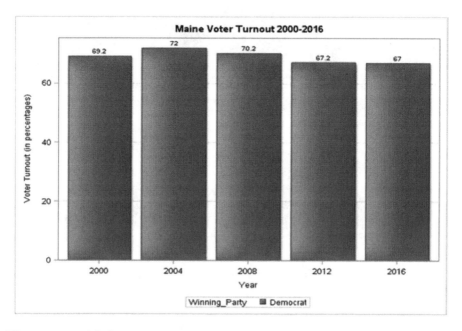

Figure 1-7. *Maine voter turnout 2000–2016*

Once more, I examined the voter turnout patterns categorized by the winning party and by age, race, and gender groups. In Figure 1-8, the letter D denotes the Democratic Party, R signifies the Republican Party, and O represents all other parties combined.

	2004			2008			2012			2016		
	D	R	O	D	R	O	D	R	O	D	R	O
Voter Turn Out	396,842	330,201	8,069	421,923	295,273	10,636	401,306	292,276	9,352	357,735	335,593	38,105
%	53.57%	44.58%	1.09%	57.71%	40.38%	1.45%	56.27%	40.98%	1.31%	47.80%	44.90%	5.10%
18 to 24	48%	50%	1%	71%	26%	3%	65%	30%		50%	42%	5%
25 to 29	48%	50%	1%	62%	36%	2%	60%	35%		46%	44%	8%
30 to 39	48%	50%	1%	60%	38%	2%	59%	36%		40%	50%	7%
40 to 49	59%	39%	1%	54%	44%	2%	51%	44%		48%	46%	6%
50 to 64	59%	39%	1%	61%	36%	3%	58%	40%		47%	48%	4%
65+	54%	45%	1%	45%	53%	2%	55%	43%		56%	39%	3%
Male	48%	49%	1%	52%	46%	2%	50%	46%		41%	52%	6%
Female	57%	42%	1%	64%	34%	2%	64%	34%		55%	39%	5%
White	53%	45%	11%	58%	40%	2%	57%	40%		47%	46%	5%
Non-White										56%	33%	10%

Figure 1-8. *Maine voter turnout according to party, age, race, and gender groupings*

Winning Candidates in 2012

I began delving into the data by constructing a histogram illustrating candidates and their respective winning percentages in the 2012 Maine election. The analysis revealed a substantial victory for Barack Obama over Mitt Romney, with other candidates receiving minimal votes, as depicted in Figure 1-9. Chapter 3 provides a detailed, step-by-step guide on replicating Figure 1-9.

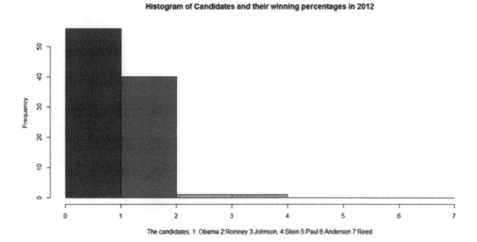

Figure 1-9. *Histogram of candidates and their winning percentages in 2012*

I then attempted to represent the data geographically, delineating it by county. Figure 1-10 illustrates Democratic-leaning counties with the darkest shade, contrasting with Republican-held regions depicted in a lighter hue from 2000 to 2016. However, it is essential to remember that geographical areas do not necessarily correlate with population density in Maine.

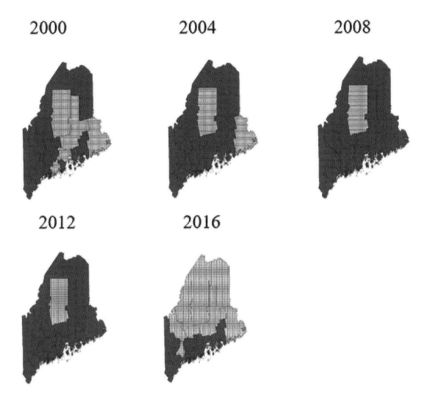

Figure 1-10. *Map of how Maine's counties voted in the presidential elections in years 2000–2016*

Categories/Issues

Each presidential election debate addresses pertinent issues and elucidates each candidate's proposed approach. These strategies can sway voter opinions and influence their decision-making based on topics. Consequently, it is imperative to examine these issues thoroughly.

Issues Facing the Country

According to CNN's report in 2016, Maine voters considered foreign policy, immigration, economy, and terrorism the most pressing national concerns. However, it is noteworthy that these issues may not directly reflect Maine's challenges. As the map indicates, Maine's local context is not significantly impacted by foreign policy matters.

Based on the race statistics, it is evident that Maine has a minimal immigrant population. A brief inquiry into any terrorist incidents in Maine yielded no results till the time of that project in 2016. However, the economy emerged as a persistent global concern. Consequently, I commenced exploring Maine-specific issues documented in the ballots.

Categories: Issues Facing Maine

I accessed a list of ballots outlined in Table 1-1 on the Ballotpedia website.

Table 1-1. *The Issues Facing Maine Residents*[1,2]

2000	2004	2008	2012	2016
Assisted death	Education	Bonds	Marriage	Marijuana
Business	Taxes	Taxes	Education	Education
Lottery	Hunt and fish	Gambling	Water	Firearms
Taxes		Water	Transportation	Economics
Suffrage		Transportation	Bond issues	Elections
LGBT				Bonds
Fishing				

[1] Data source: http://www.cnn.com/election/results/exit-polls/maine/president

[2] Data sources: https://ballotpedia.org/Maine_2000_ballot_measures
https://ballotpedia.org/Maine_2004_ballot_measures
https://ballotpedia.org/Maine_2008_ballot_measures
https://ballotpedia.org/Maine_2012_ballot_measures
https://ballotpedia.org/Maine_2016_ballot_measures

It became apparent that specific issues, such as fishing, bonds, water, education, and taxes, persist unresolved over time. This finding led me to infer that Maine grapples with water resources and taxation challenges. Once again, I underscored the importance of presidential campaign speeches addressing these specific issues by proposing viable solutions or plans. Addressing these concerns directly in speeches could sway Maine's voters.

Based on the information gathered, a review of local newspapers, and the most repeated topics discussed and unresolved on Maine ballots, I anticipate the primary five issues in Maine for 2020 to be

1. Education

2. Economics

3. Bonds

4. Healthcare

5. Taxes

Factors Affecting Maine's Economy

When discussing the economy, it is imperative to identify sources of income. According to information from the *Portland Press Herald* and revenue datasets available on the Maine.gov website[3], the income sources include

1. Agriculture

2. Forestry

3. Fishing

4. Hunting

5. Taxes

---

[3] Data source: https://www.pressherald.com/2019/03/27/maine-incomes-up-4-percent-in-2018/

6.   Management of companies and enterprises

7.   The finance and insurance sector

Comparing the Unemployment Rate of Maine to That of the Rest of the States

Donald Trump pledged to create more job opportunities and bring manufacturing companies back to the United States instead of outsourcing them. Numerous news sources suggested that his emphasis on job creation influenced the voting decisions of Rust Belt states in favor of him over Hillary Clinton. I sought to examine the unemployment rate in Maine and compare it to national figures to assess the potential resonance of Trump's promise to Maine voters. It emerged that Maine did not experience significant unemployment issues in 2016. Therefore, Trump's promise might appeal less to Maine voters than in certain other states.

Median Income for Maine

For greater precision and research objectives, I referenced the median income data for Maine. I juxtaposed it with the median income of the entire United States. As indicated in Table 1-2, from 2005 to 2015, Maine's median income remained slightly lower than the national average. Nevertheless, there was a notable increase in the three years leading to the elections.

Table 1-2. *Historical Real Median Household Income for Maine[4]*

Date	US	Maine
2018	$61,937	$55,602
2017	$61,807	$57,649
2016	$60,291	$55,542
2015	$59,116	$54,578
2014	$56,969	$52,515
2013	$56,415	$50,718
2012	$56,288	$51,180
2011	$56,507	$51,507
2010	$57,762	$52,878
2009	$58,921	$53,657
2008	$60,829	$54,459
2007	$61,601	$55,710
2006	$60,490	$54,233
2005	$59,604	$55,168

How Did Exit Polls Show the Income Factor in the 2016 Election in Maine?

Viewed from a different angle, CNN reported that exit polls indicated a strong inclination for Clinton among Maine voters, except those earning salaries between $30,000 and $50,000. Interestingly, this demographic favored Trump. However, they constituted only 19% of the surveyed sample. Consequently, the prevailing sentiment in Maine leaned toward supporting Clinton.

---

[4] Data source: http://www.deptofnumbers.com/income/maine/

Reviewing Past Elections to See If There Are Any Predictable Outcomes (Patterns)

Finally, the final phase of the project involved gathering historical data from previous elections to inform the subsequent modeling and prediction steps. As previously stated, I concentrated exclusively on Democrats and Republicans, given that other parties garnered negligible percentages in the 2012 elections.

Table 1-3. *The Historical Results of Past Presidential Elections in Maine[5,6]*

Year	D	D%	R	R%
2016	**357,735**	**47.80%**	335,593	44.90%
2012	**401,306**	**56.27%**	292,276	40.98%
2008	**421,923**	**57.71%**	295,273	40.38%
2004	**396,842**	**53.57%**	330,201	44.58%
2000	**319,951**	**49.10%**	286,616	44.00%
1996	**312,788**	**51.60%**	186,378	30.80%
1992	**263,420**	**38.80%**	206,820	30.40%
1988	243,569	43.90%	**307,131**	**55.30%**
1984	214,515	38.80%	**336,500**	**60.80%**
1980	220,974	42.30%	**238,522**	**45.60%**
1976	232,279	48.07%	**236,320**	**48.91%**

(continued)

[5] Data source: http://www.cnn.com/election/results/exit-polls/maine/president

[6] Data source: https://www.bangordailynews.com/2016/11/09/politics/elections/clinton-leads-maine-but-trump-poised-to-take-one-electoral-vote/

Table 1-3. *(continued)*

Year	D	D%	R	R%
1972	160,584	38.50%	**256,458**	**61.50%**
1968	**217,312**	**55.30%**	169,254	43.10%
1964	**62,264**	**68.84%**	118,701	31.16%
1960	**181,159**	**42.95%**	240,608	57.05%

Upon reviewing the data presented in Table 1-3, I was taken aback to discover Maine's historical political trajectory: transitioning from a predominantly Republican state to a staunchly Democratic one. Consequently, I opted to characterize Maine as a "Lean Democrat" state, given that Democrats emerged victorious in 7 out of 15 elections from 1960 to 2016. My prediction for 2020 leans toward a predominantly Democratic outcome, albeit by a narrow margin. Notably, since 1992, Maine has consistently favored the Democratic Party. However, neglecting Maine could potentially lead to a shift toward Republican allegiance. According to the *Bangor Daily News*, Clinton did not visit Maine after September 2015, opting instead to delegate surrogates. Consequently, for the first time in years, the Democratic Party lost one electoral vote to the Republican Party.

Modeling

In modeling, it is advisable to experiment with various algorithms for prediction and assess their outcomes rather than relying solely on one model. In this project, I used Monte Carlo and Bayes algorithms. Regarding statistical tests, the methods utilized included

- Histograms

- Box plots

- Proportion test

- T-test

- Decision tree

- Chi-square test

- Scatterplot and linear regression

My 2016 Predictions

Following executing the Monte Carlo algorithm in SAS University Edition using previous election outcomes, the resulting output is illustrated in Figure 1-11.

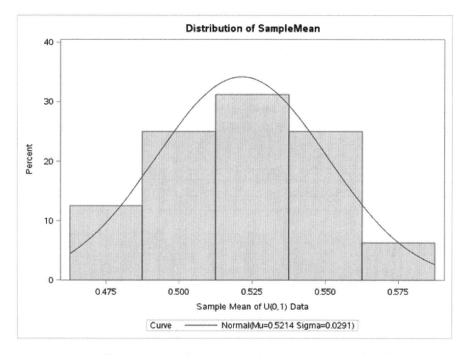

Figure 1-11. *The 2016 prediction results of Monte Carlo algorithm*

Figure 1-11 indicates that according to the simulation, Hillary Clinton was projected to win Maine with a mean percentage of 52.14%. Her victory margin[7] was only 48%. Nonetheless, the simulation accurately predicted her win in Maine, resulting in a successful outcome for this project and earning me an A in the course. Hurray!

My 2020 Predictions

Based on my prediction, the Democratic Party is expected to secure three out of Maine's four electoral votes with a majority of 50.8%, as depicted in Figure 1-12.

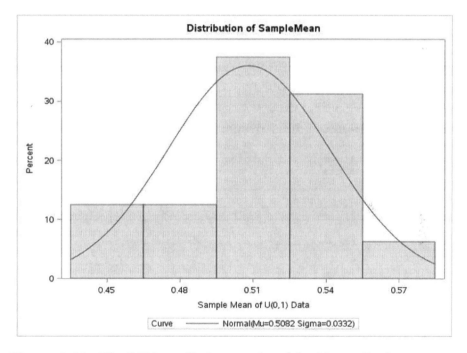

Figure 1-12. *The 2020 prediction results of the Monte Carlo algorithm*

---

[7]Data source: https://ballotpedia.org/Presidential_election_in_Maine,_2016

The probability suggests that in 2020, the Democrats are expected to edge out the Republicans in Maine by about 51% with a slim margin of approximately +3%. In 2020[8], Biden won by 53.1%, while Trump earned 44% only of the votes. Historical data from 1960 to 2016 indicates that the average percentage of Democratic support in Maine hovers around 43%, while the average percentage for Republicans is about 40%. This margin mirrors the difference between Democrats and Republicans in Maine during the 2016 election. Nonetheless, there is a risk of Maine shifting allegiance to the Republican Party if the Democratic Party fails to exercise caution.

Summary

In this chapter, our attention is directed toward extracting the key concepts from the case study of the presidential elections project in Maine, along with its accompanying charts. In the subsequent chapter, we will delve into the process of writing our initial Python program and data types.

---

[8] Data source: `https://ballotpedia.org/Presidential_election_in_ Maine,_2020`

CHAPTER 2

Getting Started

We will start this chapter by installing and taking a tour in the Jupyter Notebook interface to get more familiar with it. When studying any programming language, we always cover the following topics: the datatypes, which we will discuss in this chapter, comparison and logical operators in Chapter 5, conditional statements in Chapter 6, and loops in Chapter 5.

Installation

You can install Anaconda for free for personal and educational purposes. Anaconda is an easy-to-install manager for environment, Python distribution, and a collection of over 720 open source packages. Its main advantage is that it comes bundled with Python, Jupyter Notebook, Spyder IDE, and data analysis and machine learning libraries and packages, as shown in Figure 2-1. You can install it for free from this site:

```
https://www.anaconda.com/download/
```

For this book's example code, I used Jupyter Notebooks. However, you can copy and paste the code into your favorite Python IDE (integrated development environment). You must make sure to upload the book's datasets from the Datasets Folder and adjust the files' paths in the code accordingly. Also, you can install only the Python and Jupyter Notebook without installing the whole Anaconda. You can install Jupyter Notebook for free from this link:

```
https://jupyter.org/install
```

© Engy Fouda 2024
E. Fouda, *Learn Data Science Using Python*, https://doi.org/10.1007/979-8-8688-0935-4_2

Figure 2-1. *Anaconda*

Python on the Cloud

If you do not have Conda or Anaconda installed, you can use one of these playgrounds to access Python on the cloud:

- Kaggle: `https://www.kaggle.com/code`

- Anaconda Cloud: `https://anaconda.cloud/`

- Online Python: `www.online-python.com`

- Sololearn: `https://code.sololearn.com/python`

- Replit: `https://replit.com/languages/python3`

- Trinket: `https://trinket.io/library/trinkets/create?lang=python3`

- Google Colab: `https://colab.research.google.com`

> **Note**   All these playgrounds are free except Anaconda Cloud; it is free for a while. As the terms and conditions and pricing plans change frequently, please make sure that you read them well before signing up.

What Is Python?

Python is a programming and scripting language. It is popular, fast, and more flexible than other programming languages. The main advantage is that an enormous community contributes to it. Therefore, you will easily find batteries/packages and code samples for whatever you need for your projects.

Tour

Each place to enter text is called a cell, as shown in Figure 2-2. You can enter Python code or any text. If you want to insert any plain English text, select the markup language from the drop-down menu and choose Markdown.

To make titles in the markup cell, you type # and space before your title. When you run this cell, the Python interpreter will make it bolded and of large font, as shown in Figure 2-3.

Figure 2-2. *Cell and Markdown*

To run a cell, click the Run icon, as shown in Figure 2-3, or click Shift-Return by holding the Shift key; then, click the Return key. If you want to run all Jupyter Notebook cells, you click on the double arrows icon, as highlighted in Figure 2-3.

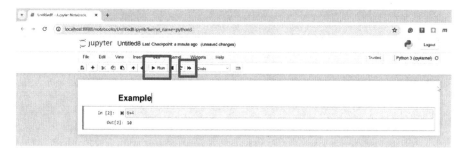

Figure 2-3. *Run and Run-All*

To delete any cell, you can click ESC, then double-click on the D key on the keyboard, or select it from the Edit menu by clicking on Delete Cell, as in Figure 2-4.

Figure 2-4. *Delete Cells*

To add any cell, click on the plus. As shown in Figure 2-5, you can move it up and down by clicking on the icons in the toolbar. You can also move it from the Insert menu as well.

Figure 2-5. *Adding Cells*

Again, it depends on your installed version; the interface might vary. In this book, the version is Jupyter Notebook server 6.4.5. The server running on this version of Python is Python 3.9.7 (default, Sep 16 2021, 16:59:28) [MSC v.1916 64 bit (AMD64)]. The kernel information is Python 3.9.7 (default, Sep 16 2021, 16:59:28) [MSC v.1916 64 bit (AMD64)].

Datatypes

There are different ways to classify the datatypes in Python. Usually, it will be in the following six categories:

1. Numeric Types

   a. int: Used for whole numbers, e.g., 42

   b. float: Used for decimal numbers, e.g., 3.14

   c. complex: Used for complex numbers, e.g., 3+4j

2. Sequence Types

   a. str: Used for text, e.g., "hello"

   b. list: Used for ordered collections of items, e.g., [1, 2, 3]

   c. tuple: Used for ordered, immutable collections of items, e.g., (1, 2, 3)

3. Mapping Types

   a. dict: Dictionary type used for key/value pairs, e.g., {'name': 'Alice', 'age': 25}

4. Set Types

   a. set: Used for unordered collections of unique items, e.g., {1, 2, 3}

   b. frozenset: Immutable set type, e.g., frozenset([1, 2, 3])

5. Boolean Types

   a. bool: Boolean type, used for true/false values, e.g., True, False

6. Binary Types

    a. bytes: Immutable sequence of bytes, e.g., `b'hello'`

    b. bytearray: Mutable sequence of bytes, e.g., `bytearray(b'hello')`

    c. memoryview: Memory view object, e.g., `memoryview(b'hello')`

Although this classification is standard, I prefer classifying it into scalar/basic, container, and advanced. Therefore, the next section of the typecasting will follow this classification.

1. Scalar types are simple datatypes that hold a single value. It includes int, float, bool, complex, and str.

2. Container types are data structures with multiple values, which can be of different datatypes. This class includes list, tuple, dict, set, and frozenset.

3. Advanced types typically come from external libraries for more complex data structures and operations.

    a. Pandas DataFrame: Two-dimensional labeled data structure with columns of potentially several types used for data analysis, as in Listing 2-1. The whole code is in Chapter 2.ipynb in the Example Code Folder.

Listing 2-1. Pandas DataFrame

```
import pandas as pd
df = pd.DataFrame({'A': [1, 2, 3], 'B': [4, 5, 6]})
print(df)
```

The output is as in Figure 2-6. The first column on the left is not a component of the DataFrame; it is an index that serves as an identifier for the rows.

```
      A  B
0  1  4
1  2  5
2  3  6
```

Figure 2-6. Pandas DataFrame is like a datasheet which consists of columns and rows

b. Numpy array: N-dimensional array object for numerical operations, as in Listing 2-2.

Listing 2-2. Numpy

```
import numpy as np
arr = np.array([1, 2, 3])
print(arr)
```

The output is in Figure 2-7.

$$[1\ 2\ 3]$$

Figure 2-7. One-dimensional Numpy array

c. Pandas Series: One-dimensional labeled array capable of holding any datatype, as in Listing 2-3.

Listing 2-3. Pandas Series

```
import pandas as pd
s = pd.Series([1, 2, 3])
print(s)
```

The output is in Figure 2-8. We use the Pandas Series to store one column from a Pandas DataFrame, as we are going to see in many of the book's examples.

```
0      1
1      2
2      3
dtype: int64
```

Figure 2-8. *Pandas Series*

We will discuss Pandas and Numpy extensively later in this chapter and learn how to use them effectively throughout the book.

Typecasting

Typecasting is converting from one datatype to another. Here are examples within these categories:

Scalar Type Casting

1. int(): Convert to an integer, as in Listing 2-4.

Listing 2-4. Convert to integer

```
x = int(3.14)  # x is now 3
print(x)
```

2. float(): Convert to float, as in Listing 2-5.

Listing 2-5. Convert to float

```
x = float(3) # x is now 3.0
print(x)
```

3. str(): Convert to string, as in Listing 2-6.

Listing 2-6. Convert to string

```
x = str(3.14) # x is now '3.14'
print(x)
```

Container Type Casting

1. list(): Convert to list, as in Listing 2-7.

Listing 2-7. Convert to list

```
x = list((1, 2, 3))  # x is now [1, 2, 3]
print(x)
```

2. tuple(): Convert to tuple, as in Listing 2-8.

Listing 2-8. Convert to tuple

```
x = tuple([1, 2, 3])  # x is now (1, 2, 3)
print(x)
```

3. set(): Convert to set, as in Listing 2-9.

Listing 2-9. Convert to set

```
x = set([1, 2, 2, 3])  # x is now {1, 2, 3}
print(x)
```

4. dict(): Convert to dictionary (usually from a list of key/value pairs), as in Listing 2-10.

Listing 2-10. Convert to dict

```
x = dict([('a', 1), ('b', 2)])  # x is now {'a': 1, 'b': 2}
print(x)
```

Advanced Type Casting

1. to_numpy(): Convert a Pandas DataFrame or Series to a Numpy array, as in Listing 2-11, using arr = df.to_numpy()

Listing 2-11. Convert to Numpy array

```
import pandas as pd
import numpy as np

# Step 1: Create a pandas DataFrame
data = {
    'A': [1, 2, 3, 4],
    'B': [5, 6, 7, 8],
    'C': [9, 10, 11, 12]
}
df = pd.DataFrame(data)
print("Original DataFrame:\n", df)

# Step 2: Convert the DataFrame to a numpy array
numpy_array = df.to_numpy()
print("\nConverted to numpy array:\n", numpy_array)
```

The output is in Figure 2-9.

```
Original DataFrame:
     A  B   C
0    1  5   9
1    2  6  10
2    3  7  11
3    4  8  12

Converted to numpy array:
 [[ 1  5  9]
  [ 2  6 10]
  [ 3  7 11]
  [ 4  8 12]]
```

Figure 2-9. From Pandas DataFrame to Numpy array

2. pd.DataFrame(): Create a Pandas DataFrame from a Numpy array, as in Listing 2-12, using
   df= pd.DataFrame(np_array)

Listing 2-12. Convert to Pandas Dataframe

```python
import numpy as np
import pandas as pd

# Step 1: Create a numpy array
numpy_array = np.array([[1, 2, 3], [4, 5, 6], [7, 8, 9]])
print("Original numpy array:\n", numpy_array)

# Step 2: Convert the numpy array to a pandas DataFrame
df = pd.DataFrame(numpy_array, columns=['Column1', 'Column2',
'Column3'])
print("\nConverted to DataFrame:\n", df)
```

The output is in Figure 2-10.

```
Original numpy array:
 [[1 2 3]
 [4 5 6]
 [7 8 9]]

Converted to DataFrame:
     Column1  Column2  Column3
0          1        2        3
1          4        5        6
2          7        8        9
```

Figure 2-10. *From Numpy array to Pandas DataFrame*

Pandas

Throughout this book, we will use Pandas and Numpy extensively. Pandas is a powerful and popular Python library for data manipulation and analysis. It provides easy-to-use data structures and functions designed to work with structured data seamlessly. One of its notable features is the ability to present data in a human-readable table format while also allowing numerical interpretation for extensive computations. Think of Pandas as a supercharged version of Excel within Python, allowing you to handle and analyze data more efficiently and programmatically.

Why Use Pandas?

- Easy Data Handling: Pandas simplifies tasks like loading data from files, cleaning messy data, and performing calculations.

- Data Analysis: With Pandas, you can explore large datasets quickly, filter out the information you need, and gain insights.

- Versatility: Pandas can work with different datatypes, including numbers, text, and time series data.

Key Concepts

- Series: A Series is a one-dimensional array-like object that can hold various datatypes, including numbers, strings, and dates. Think of it as a single column in a spreadsheet, as shown in Listing 2-13.

Listing 2-13. Pandas Series

```
import pandas as pd
df = pd.DataFrame({'A': [1, 2, 3], 'B': [4, 5, 6]})
series=df['A']
print(series)
```

The output is in Figure 2-11.

```
0    1
1    2
2    3
Name: A, dtype: int64
```

Figure 2-11. *Getting a Series from a Pandas DataFrame*

- DataFrame: A DataFrame is a two-dimensional table similar to a spreadsheet or a SQL table. It consists of rows and columns, where each column can have a different datatype, as in Listing 2-14.

Listing 2-14. Pandas DataFrame

```
data = {
        'Name': ['Alice', 'Bob', 'Charlie'],
        'Age': [25, 30, 35],
        'City': ['New York', 'Los Angeles', 'Chicago']
    }
df = pd.DataFrame(data)
print(df)
```

The output is in Figure 2-12.

```
      Name  Age         City
0    Alice   25     New York
1      Bob   30  Los Angeles
2  Charlie   35      Chicago
```

Figure 2-12. *Creating a Pandas DataFrame*

Basic Operations with Pandas

- Loading Data: You can load data from various sources, such as CSV files, Excel files, and databases. In the coming chapters, we will see how to load multiple file types.

- Inspecting Data: Check the structure and summary of the data. To summarize a DataFrame, use the df.info() method. To make a statistical summary, use the df.describe() method, as we will see in Chapter 3.

- Filtering Data: In the next chapter, Chapter 3, we will see how to select specific rows and columns based on how we want to manipulate and analyze our data to generate the required chart for visualizations. Through the rest of the book, we keep using the same techniques for further data science tasks.

- Modifying Data: Add, remove, or update columns. In Chapter 6, we will discuss how to do each of these tasks in full detail.

- Managing Missing Data: Identify and handle missing values. Again, in Chapter 6, we will see how to use `df.dropna()` to remove rows with missing values, `df.fillna(0)` to replace missing values with 0, and other methods.

- Indexing in Pandas DataFrame: Access a column by name, access a row by index, use `df.at[row number, 'column name']` property to access a specific element by row and column, and use `df.loc[row start:row end, 'column start':'column end']` to slice both rows and columns. We will have more examples demonstrating this topic in Chapter 6.

Numpy

Numpy, short for "numerical Python," is a foundational Python library for numerical computing. It supports arrays, matrices, and high-level mathematical functions to operate on these data structures. Numpy is the backbone of many other data science libraries, including Pandas.

We frequently transfer data from our Pandas DataFrame to Numpy arrays. Pandas DataFrames are beneficial because they include column names and other text data, making them easy for humans to read. However, Numpy arrays are the most efficient for computers to perform calculations.

Why Use Numpy?

- Performance: Numpy operations are implemented in C, making them extremely fast and efficient for large-scale computations.

- Ease of Use: Numpy makes it easy to perform complex mathematical operations on arrays and matrices with simple syntax.

- Flexibility: Numpy arrays can handle various datatypes and are more versatile than Python lists for numerical computations.

Key Concepts

- Array: A Numpy array is a grid of values, all the same type, as in Listing 2-15. A tuple of nonnegative integers indexes arrays.

Listing 2-15. Numpy array

```
import numpy as np
array = np.array([1, 2, 3, 4, 5])
print(array)
```

The output is in Figure 2-13.

$$[1\ 2\ 3\ 4\ 5]$$

Figure 2-13. *Creating a Numpy array*

- Matrix: A matrix is a two-dimensional array. Numpy can handle both one-dimensional and two-dimensional arrays effortlessly, as in Listing 2-16.

Listing 2-16. Numpy matrix

```
matrix = np.array([[1, 2, 3], [4, 5, 6], [7, 8, 9]])
print(matrix)
```

The output is in Figure 2-14.

$$[[1\ 2\ 3]$$
$$[4\ 5\ 6]$$
$$[7\ 8\ 9]]$$

Figure 2-14. *Creating a Numpy matrix*

Basic Operations with Numpy

- Creating Arrays: Numpy provides several functions to create arrays quickly, as in Listing 2-17.

Listing 2-17. Creating Numpy arrays

```
zeros = np.zeros((3, 3))  # Create a 3x3 matrix filled
                                with zeros
ones = np.ones((2, 2))  # Create a 2x2 matrix filled with ones
print(zeros)
print(ones)
```

The output is in Figure 2-15.

```
[[0. 0. 0.]
 [0. 0. 0.]
 [0. 0. 0.]]
[[1. 1.]
 [1. 1.]]
```

Figure 2-15. *Creating two Numpy matrices, one is 3×3 filled with zeros and a smaller one of 2×2 matrix filled with ones*

- Array Operations: Perform element-wise operations on arrays, as in Listing 2-18.

Listing 2-18. Array operations

```
array1 = np.array([1, 2, 3])
array2 = np.array([4, 5, 6])
sum_array = array1 + array2  # Element-wise addition
print("array1=",array1)
print("array2=",array2)
print("sum_array=",sum_array)
```

The output is in Figure 2-16. Performing arithmetic operations using Numpy is easy.

```
array1= [1 2 3]
array2= [4 5 6]
sum_array= [5 7 9]
```

Figure 2-16. *Summation of two Numpy arrays*

- Mathematical Functions: Numpy provides a range of mathematical functions that can be applied to arrays, as in Listing 2-19.

41

Listing 2-19. Mathematical functions

```
array = np.array([1, 2, 3, 4, 5])
print("array=",array)
print("square root=",np.sqrt(array))  # Square root of
                                         each element
print("mean=",np.mean(array)) # Mean of the array
```

The output is in Figure 2-17. Performing mathematical operations using Numpy is easy.

```
array= [1 2 3 4 5]
square root= [1.        1.41421356 1.73205081 2.        2.23606798]
mean= 3.0
```

Figure 2-17. *Square root and mean of a Numpy array*

- Indexing and Slicing: Access and modify specific elements or subarrays, as in Listing 2-20.

Listing 2-20. Indexing and slicing

```
array = np.array([1, 2, 3, 4, 5])
print("array=",array)
print("second element=array[1]=",array[1]) # Access the
                                         second element
# Access elements from index 1 to 3
print("From second to fourth elements=",  array[1:4])
array[2] = 10  # Modify the third element
print("new array after changing third element value=",array)
```

The output is in Figure 2-18. We will have plenty of practical real-world hands-on-labs to demonstrate useful indexing and slicing in the data science process through the book, especially in Chapter 6.

```
array= [1 2 3 4 5]
second element=array[1]= 2
From second to fourth elements= [2 3 4]
new array after changing third element value= [ 1  2 10  4  5]
```

Figure 2-18. *Indexing and slicing the Numpy array*

- Reshaping Arrays: Change the shape of an array without changing its data, as in Listing 2-21.

Listing 2-21. Reshaping arrays

```
array = np.array([[1, 2, 3], [4, 5, 6]]) # 3x3 array
reshaped = array.reshape((3, 2))  # Reshape to a 3x2 array
print(reshaped)
```

The output is in Figure 2-19. We first create a 3×3 Numpy matrix; then, we reshaped it using the .reshape() method to modify the matrix dimensions to 3×2. Plenty of data science and machine learning functions in Python need to reshape the matrices to prepare the data for processing.

```
[[1 2]
 [3 4]
 [5 6]]
```

Figure 2-19. *Reshaping the matrix from 3×3 to 3×2 without losing and changing data*

Summary

This chapter represents the foundation of our journey in learning Python. It starts with the installation and Python Cloud Playgrounds and then explains the interface and the essential components of a Jupyter Notebook. It also describes the datatypes and typecasting.

Moreover, it explains Pandas and Numpy, which are crucial tools for data manipulation and numerical computations in Python. Pandas provide high-level data structures and functions for working with structured data, making it easy to analyze and visualize data.

Numpy, on the other hand, offers powerful capabilities for numerical operations on arrays and matrices, providing the foundation for efficient scientific computing. Together, these libraries form a powerful base for delving into data. In the next chapter, we shall dig deeper into the data visualization.

CHAPTER 3

Data Visualizations

There is an adage that says, "A picture is worth a thousand words." Visualization is a tedious and harsh task. Luckily, Python has plenty of batteries that contain functions with options to make expressive graphs. This chapter uses big datasets that are available in the public domain. We shall follow the data science process mentioned in Chapter 1, where we start every section with a question and seek its answer through a graph.

This chapter covers various essential charts, such as scatterplots, histograms, series plots, bar charts and sorted bar charts, and bubble plots.

Scatterplot

Scatterplots are used to show the interdependence between variables by describing the scatterplots' direction, strength, and linearity. Moreover, they can easily display the outliers.

For this example, our question is: Do the salaries in the City of Seattle depend on the age range of the employees, or is there no relationship?

You will find the data file in the Datasets Folder, or you can download the CSV file from Data.Gov project at the following link: `https://web.archive.org/web/20170116035611/http://catalog.data.gov/dataset/city-of-seattle-wages-comparison-by-gender-average-hourly-wage-by-age-353b2`.

© Engy Fouda 2024
E. Fouda, *Learn Data Science Using Python*, https://doi.org/10.1007/979-8-8688-0935-4_3

The dataset is in the Datasets Folder with the name: City_of_Seattle_
Wages__Comparison_by_Gender_-_Average_Hourly_Wage_by_Age.csv.
The whole code is in Listing 3-1 in Chapter 3.ipynb in the Example Code
Folder. However, in the Jupyter Notebook, the code will be divided in cells
matching the steps below.

There are multiple libraries that you can use for visualizations. In this
example, we use the Matplotlib. The steps are as follows:

1. Load the libraries.

2. Load the data and explore it.

3. Create a scatterplot.

4. Enhance the plot by

    a. Removing some unneeded values

    b. Adding a title

    c. Resizing the plot

    d. Giving labels to the x axis and the y axis

Listing 3-1-1. Import the required libraries

```
import pandas as pd
import matplotlib.pyplot as plt
import numpy as np
```

In Listing 3-1-1, we import the required libraries. We start by Pandas
modules as pd. Pandas are one of the reasons Python is so popular. It helps
with reading and manipulating data by loading the data; viewing it as a
table, which is human readable; and doing lots of computations with it.
The table of data is called a DataFrame, and the shortened convention is
df. Pandas DataFrames have column names and other text data that makes
it user-friendly.

Then, we load the `matplotlib.pyplot` batteries/libraries where all the visualization functions are. It is a convention to nickname it as `plt`.

The last library to load is `numpy`, which is a Python package for manipulating lists and tables of numerical data. The computers find it easier and faster to perform the statistical calculations as numpy arrays. However, they are not very human friendly. Therefore, we often convert the data from Pandas DataFrame to Numpy arrays.

Note   We will frequently use these three libraries: Pandas, Matplotlib, and Numpy, for the examples in this book.

Listing 3-1-2. Load the dataset, and explore it

```
df=pd.read_csv('../Datasets/Chapter 3/City_of_Seattle_Wages__
Comparison_by_Gender_-_Average_Hourly_Wage_by_Age.csv')

print(df.head())
```

In Listing 3-1-2, we pull the data into Pandas to view it as a DataFrame. The read_csv() function takes a file in csv format and converts it to a Pandas DataFrame, called df. Then, the head method returns the first five rows of the DataFrame.

The output of the head is as in Figure 3-1.

```
print(df.head())
```

	AGE RANGE	Average of FEMALE HOURLY RATE	Count of FEMALE EMPLID \
0	20 - 25	22.92	26
1	26 - 30	29.90	163
2	31 - 35	31.43	292
3	36 - 40	34.90	376
4	41 - 45	36.05	498

	Average of MALE HOURLY RATE	Count of MALE EMPLID \
0	26.30	63
1	31.64	325
2	35.54	530
3	37.10	726
4	39.28	985

	Total Average of HOURLY RATE	Total Count of EMPLID	Female to male % rate
0	25.32	89	87.16
1	31.06	488	94.50
2	34.08	822	88.43
3	36.35	1102	94.07
4	38.19	1483	91.78

Figure 3-1. *Output of df.head() method*

From Figure 3-1, we find that all eight columns are numeric except the AGE RANGE. It is of a string datatype.

Let us explore the dataset more by using the describe() method. It returns the summary statistics report of the dataset, as in Listing 3-1-3.

Listing 3-1-3. The describe method computes the summary statistics

```
print(df.describe())
```

```
            Average of FEMALE HOURLY RATE  Count of FEMALE EMPLID   \
count                          13.000000                 13.000000
mean                           31.956923                553.846154
std                             6.236531                943.730439
min                            15.760000                  1.000000
25%                            30.800000                 75.000000
50%                            34.900000                369.000000
75%                            36.050000                562.000000
max                            37.190000               3600.000000

            Average of MALE HOURLY RATE  Count of MALE EMPLID   \
count                        13.000000             13.000000
mean                         35.443846            966.923077
std                           7.942968           1645.390656
min                          13.430000              1.000000
25%                          34.520000            131.000000
50%                          39.040000            530.000000
75%                          40.660000            985.000000
max                          41.390000           6285.000000

            Total Average of HOURLY RATE  Total Count of EMPLID   \
count                          13.000000              13.000000
mean                           34.220769            1520.769231
std                             7.164340            2588.651166
min                            14.590000               2.000000
25%                            33.420000             206.000000
50%                            37.700000             890.000000
75%                            38.280000            1483.000000
max                            39.770000            9885.000000

            Female to male % rate
count                  13.000000
mean                   91.560000
std                     8.493014
min                    81.050000
25%                    88.430000
50%                    89.920000
75%                    91.780000
max                   117.350000
```

Figure 3-2. *Figure 3-2 shows the output of the describe() method*

Figure 3-2 shows a few statistics for each column. Note that it only gives statistics for the numerical columns. Here are the statistics definitions:

- Count: This is the count of non-missed rows.

- Mean: The average.

- Std: This is short for standard deviation. This is a measure of how dispersed the data is.

- Min: The smallest value.

- 25%: The 25th percentile.

- 50%: The 50th percentile, also known as the median.

- 75%: The 75th percentile.

- Max: The largest value.

We use the Pandas describe() method to start building some intuition about our data.

Listing 3-1-4. The initial scatterplot

```
plt.figure(figsize=(12, 6))
plt.scatter(df['AGE RANGE'], df['Total Average of
HOURLY RATE'])
plt.xlabel('AGE RANGE')
plt.ylabel('Total Average of HOURLY RATE')
```

To avoid having the plots being drawn over each other, we use the figure method, where we set up the plot dimensions as well, as in Listing 3-1-4. Then, we use the scatter method and pass to it the two variables we want to plot. For enhancing the diagram, we set the label of the x and y axes using the xlabel and ylabel methods.

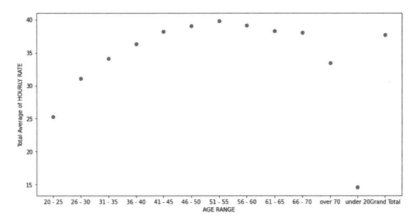

Figure 3-3. *Output of the scatterplot*

In Figure 3-3, the last two values, the AGE RANGE are not needed and give the wrong impression about the relationship between the two variables. Hence, we shall filter and remove them from the data, as in Listing 3-1-5.

Listing 3-1-5. Filter the last two values in the AGE RANGE

```
plt.figure(figsize=(12, 6))
df1=df['AGE RANGE']
df2=df['Total Average of HOURLY RATE']
plt.scatter(df1[:-2],df2[:-2] )
plt.xlabel('AGE RANGE')
plt.ylabel('Total Average of HOURLY RATE')
plt.title('Interdependence between Age Range and Total Average
of HOURLY RATE')
```

From the final scatterplot in Figure 3-4, we can assure that there is an interdependence between the AGE RANGE and Total Average of HOURLY RATE in Seattle. The relationship is not linear; it is curvilinear.

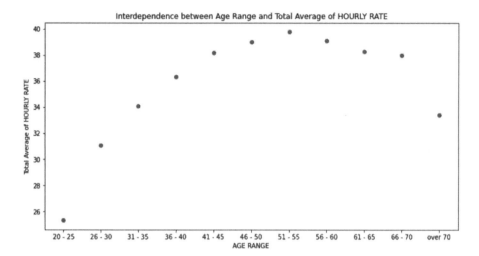

Figure 3-4. *The final scatterplot*

Scatterplot Relationships

Figure 3-5 displays different examples of scatterplot relationships. To describe the scatterplot, you should state if it is a linear relationship or curvilinear as we have seen in the previous example.

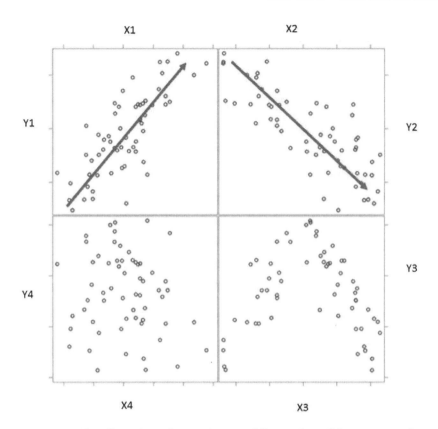

Figure 3-5. *The direction, intensity, and linearity of the scatterplot*

If the interdependency is linear, look at its direction to know if it is positive or negative. In Figure 3-5, up-left, X1-Y1, the relationship is positive, while the plot up-right, X2-Y2, is a negative relationship. In Figure 3-8, down-right, X3-Y3, there is no direction because the relationship is not linear but curvilinear. In Figure 3-5, down left, X4-Y4, there is no relationship as the points are scattered irregularly.

Finally, describe the strength of the relationship. If the points are so close to the line/form, then there is high interdependency, and the relationship is strong. Otherwise, it is a weak relationship.

Plotting More Than One Scatterplot in the Same Image

We plot more than one scatterplot in the same image to compare values. For this example, we have the number of registered voters in every party and will plot them to compare which party has more registered voters in each of Maine's counties.

The dataset MAINECOUNTIESPARTIES.xlsx in the Datasets Folder has three columns: the counties and the registered voters per party, as in Table 3-1.

Table 3-1. *Maine Registered Voters for Political Parties in Every County*

Counties	Republican	Democratic
Androscoggin	28189	22975
Aroostook	19419	13377
Cumberland	57697	102935
Franklin	7900	7001
Hancock	13682	16107
Kennebec	29296	31753
Knox	9148	12440
Lincoln	9727	10241
Oxford	12172	16214
Penobscot	41601	32832
Piscataquis	5403	3098
Sagadahoc	9304	10679
Somerset	14998	9092
Waldo	10378	10442
Washington	9037	6358
York	50388	55828

As always, we read the dataset and explore it to verify that it has been read correctly using the head() method, as in Listing 3-2-1.

Listing 3-2-1. Reading and exploring the dataset

```
df=pd.read_excel('../Datasets/Chapter 3/
MAINECOUNTIESPARTIES.xlsx')
print(df.head())
```

The output is as in Figure 3-6 showing the first five rows of the dataset.

```
     counties  Republican  Democratic
0  Androscoggin       28189       22975
1     Aroostook       19419       13377
2    Cumberland       57697      102935
3      Franklin        7900        7001
4       Hancock       13682       16107
```

Figure 3-6. *Output of the head() method*

The code plots two overlapped scatterplots where both have counties on their x axis, while the y axis in the first scatterplot is the number of the Democrats, and in the second one is the number of the Republicans. To change the colors to distinguish between each one of them, we set the color option, as in Listing 3-2-2, using c=['r']. Moreover, we change the marker of the Republican Party points only to distinguish between the points even in grayscale using marker option to make it as a triangle using marker='^' option to the scatter function. We changed the label of the y axis to "Republican:Red Triangles Democrat:Blue Circles." Finally, we set the title of the plot.

Listing 3-2-2. Comparing the Republican registered voters to the Democratic ones in Maine's counties

```
plt.figure(figsize=(18, 6))
plt.scatter(df['counties'],df['Democratic'], c=['b'])
plt.scatter(df['counties'],df['Republican'] ,
marker='^',c=['r'])

plt.xlabel('counties')
plt.ylabel('Republican:Red Triangles Democrat:Blue Circles')
plt.title('Maine Counties and Parties')
```

The output of the above code is as in Figure 3-7.

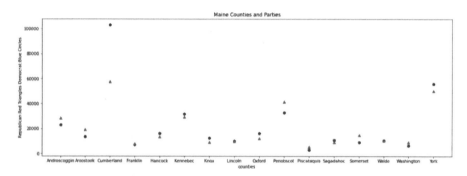

Figure 3-7. *Scatterplot of Maine counties and parties*

From Figure 3-7, we can say that Cumberland County has the highest Democratic voters and the highest registered voters in general. It might imply that it has the highest population as well, but we cannot ensure that and will need further verification. However, we are sure that the difference between the Democrats and Republicans is enormous there. Other than this county, Democrats are more than the Republicans in 8 of the remaining 15 counties with trivial differences. Hence, Democrats must give more care to Maine, so it does not turn back to red.

Moreover, if the Democratic party plans to do a rally in Maine, it should do it in Cumberland. On the other hand, the Republic party should do it in Penobscot.

Similarly, you can do the same exercise with your state and get ready for the 2024 Presidential Elections with your analysis and predictions.

Histogram

The histogram task creates a chart that displays the frequency distribution of a numeric variable. For this example, we shall see how annual wages of employees are being distributed in March 2018 at the City of Charlotte, North Carolina.

You will find the data file in the Datasets Folder, or you can download the CSV file from the following link: `https://data.amerigeoss.org/ dataset/city-employee-salaries-march-2018/resource/264fa25e- 93cb-46b8-8edd-24ad224f3b74?view_id=4c9ee967-`a039-4d72- ad78-2f556909b207`.

As always, load the CSV file to a DataFrame using the Pandas `read_ csv()` method. Then, explore it using the `head()` method to verify that it has been read correctly, as in Listing 3-3-1.

Listing 3-3-1. Reading and exploring the dataset

```
df=pd.read_csv('../Datasets/Chapter 3/City_Employee_Salaries_
March_2018.csv')
print(df.head())
```

The output is as in Figure 3-8 showing the first five rows of the dataset. We have nine columns. However, we are interested in one column only, `Annual_Rt`, to plot our histogram.

```
                  Name    Unit                  Dept         Job_Title  \
0       Jones,Marcus D.   CTMGR   City Manager's Office      City Manager
1      Hagemann,Robert E  ATTOR           City Attorney      City Attorney
2       Lewis Jr,John M    CATS               Executive   Transit  Director
3         Cagle,Brent D    AVIA      Admin - Executive    Aviation Director
4  Joy-Hogg,Sabrina Beena  CTMGR   City Manager's Office  Deputy City Manager

   Annual_Rt  Hrly_Rate Full_Part Reg_Temp  FID
0  318000.00     152.88        F        R    1
1  248491.82     119.47        F        R    2
2  245924.30     118.23        F        R    3
3  236042.01     113.48        F        R    4
4  225500.00     108.41        F        R    5
```

Figure 3-8. *Output of the head() method*

In Listing 3-3-2, we again start by using the .figure() method to draw the histogram in a new plot and do not override the histogram over the scatterplots of the previous section. Also, we set the plot dimension using the figsize in the .figure() method. Then, we select the Annual_RT as the variable in the .hist() method. Finally, we label the x axis and set the plot title.

Listing 3-3-2. Select the analysis variable to plot its histogram

```
plt.figure(figsize=(18, 10))
plt.hist(df['Annual_Rt'])
plt.xlabel('Annual_Rt')

plt.title('The Distribution of The Annual Rate of Charlotte
City Employees')
```

Figure 3-9 shows the histogram plot. However, we need to enhance the appearance of this plot. It is a right skewed curve, and there are some clear outliers. Also, we can resize the bins to have a more normally distributed curve. The last step will be adding a normally distributed curve over the histogram. We shall do all these steps in Listing 3-3-3.

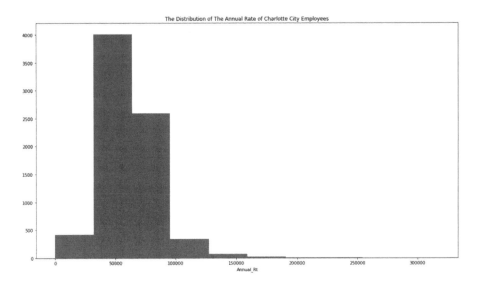

Figure 3-9. *Histogram of Charlotte City employees*

Listing 3-3-3 starts with importing the norm from spicy.stats. This library is for plotting the normal distribution curve; it has functions for working with normal (Gaussian) distributions.

Then, we generate an array of values for the x axis ranging from 0 to 14000 with step=1. We set the maximum annual salary rate to 140000 and will ignore any value larger than that by setting a mask over the Annual_RT series.

Afterward, we plot the histogram using the new masked series, change the bins to 10, and normalize the histogram to draw the density and return a probability density. Next is plotting the normal distribution with mean and standard deviation calculated from the data. The final changes are giving a label to the x axis, setting the title, setting the legend, and showing all that.

Listing 3-3-3. Enhancing the appearance of the histogram

```
from scipy.stats import norm

plt.figure(figsize=(15, 10))

#Generate data for x-axis ranging from 0 to 14000 with 1 step
x = np.arange(0, 140000, 1)

#Set the maximum salary to 140000
#Creates a new series for the 'Annual_Rt' from the DataFrame df
ar = df['Annual_Rt']
#Creates a boolean mask where values in 'Annual_Rt' are less
than 140000
mask = ar < 140000
#Applies the mask to filter the data
df1 = ar[mask]

# Plot the histogram
#plt.hist: Plots a histogram of the filtered data (df1)
#bins=10: Specifies the number of bins for the histogram
#density=True: Normalizes the histogram to represent a
probability density
#alpha=0.6: Sets the transparency of the bars
#color=b: Sets the color of the bars to blue
plt.hist(df1, bins=10, density=True, alpha=0.6, color='b')

# Plot the normal distribution with mean and standard deviation
calculated from the data
#Calculates the mean (mu) and standard deviation (std) of the
filtered data #using a normal distribution fit
mu, std = norm.fit(df1)
```

#Plots the normal distribution curve using the calculated mean and standard #deviation. The curve is in red and labeled as Normal Distribution

```
plt.plot(x, norm.pdf(x, mu, std), 'r-', label='Normal
Distribution')
```

#Sets the label for the x-axis

```
plt.xlabel('Annual_Rt')
```

#Sets the title of the plot

```
plt.title('The Distribution of The Annual Rate of Charlotte
City Employees')
```

#Displays a legend to identify the histogram and the normal distribution curve

```
plt.legend()
```

#Displays the plot

```
plt.show()
```

The output of Listing 3-3-3 is Figure 3-10 showing the normal distribution curve over the histogram after masking the salaries and setting their maximum.

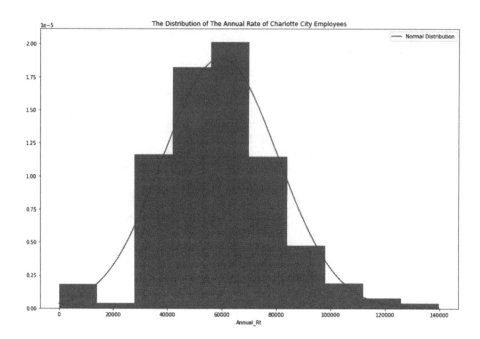

Figure 3-10. *The final histogram*

Now, we have a more normal distribution curve than before where we had a right-skewed curve.

Series Plot

The series plot displays the values against time. In this example, let us plot stock trends of prices over time.

In this exercise, we do not have a dataset to load. Rather, we will use the Yahoo Finance library, yfinance, to load the stocks data of any company. First, as in Listing 3-4-1, we will load Apple Inc. stocks' historical data. Then, in Listing 3-4-2, we will load multiple companies and plot them vs. each other to compare values. The companies are Apple (AAPL), Intel (INTC), Microsoft (MSFT), and ASML Holding (ASML).

Listing 3-4-1. Series plot to show Apple Inc. closing prices over time

```python
# This library is used to fetch financial data, including
historical stock #prices
import yfinance as yf
#matplotlib is used for creating plots and visualizations
import matplotlib.pyplot as plt

# Define the stock symbol and time range
stock_symbol = "AAPL"  # Example: Apple Inc.
start_date = "2022-01-01"
end_date = "2023-01-01"

# Download historical stock data
stock_data = yf.download(stock_symbol, start=start_date,
end=end_date)

# Plot the closing prices over time
plt.figure(figsize=(15, 8))
plt.plot(stock_data['Close'], label=f"{stock_symbol} Closing
Prices", color='blue')
plt.title(f"{stock_symbol} Closing Prices Over Time")
plt.xlabel("Date")
plt.ylabel("Closing Price")
plt.legend()
plt.grid(True)
plt.show()
```

Now, let us explain the Listing 3-4-1 code line by line. As we mentioned, start by loading the required libraries. Then, we set the stock symbol of the company and the start and the end dates. We feed that to the yf.download() method to download the historical data for the specified stock symbols. Then, we use the plt.plot() method to plot the series plot by setting the color. Finally, as usual, we set the plot title and the labels of the x and y axes, show grid, and display the plot with all these settings. The output is as in Figure 3-11.

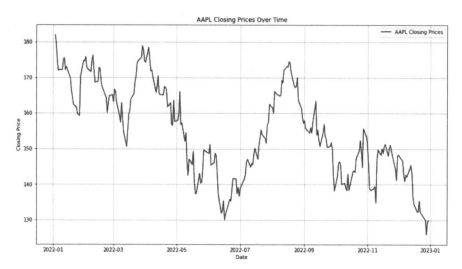

Figure 3-11. *Series plot output*

We can add more companies' stocks by adding their symbols as a list in the stock symbol, as follows: stock_symbols = ["AAPL", "INTC", "MSFT", "ASML"], as in Listing 3-4-2.

Moreover, in Listing 3-4-2, we loop over the companies' list, stock_symbols, to draw their series plots in one figure.

Listing 3-4-2. Series plot showing multiple companies' closing prices over time aggregated together

```
# yfinance is used to fetch financial data, including
historical stock #prices
import yfinance as yf
#matplotlib.pyplot used for creating plots and visualizations
import matplotlib.pyplot as plt
```

```
#stock_symbols are a list containing the stock symbols for
Apple (AAPL), #Intel (INTC), Microsoft (MSFT), and ASML
Holding (ASML)
stock_symbols = ["AAPL", "INTC", "MSFT", "ASML"]
#The date range to fetch historical stock data
start_date = "2021-01-01"
end_date = "2023-01-01"

#Download historical stock data for each stock
stock_data = yf.download(stock_symbols, start=start_date,
end=end_date)['Close']

#A loop is used to iterate over each stock symbol, and the
closing prices #for each stock are plotted. All in the
same figure
plt.figure(figsize=(15, 8))

for symbol in stock_symbols:
    plt.plot(stock_data[symbol], label=f"{symbol} Closing Prices")

plt.title("Closing Prices Over Time for Selected Stocks")
plt.xlabel("Date")
plt.ylabel("Closing Price")
# Displays a legend to identify each stock
plt.legend()
# Adds a grid to the plot for better readability
plt.grid(True)
#Show the plot using the above settings
plt.show()
```

The output of Listing 3-4-2 is Figure 3-12. This code fetches historical closing prices for the specified stocks and plots them on the same graph for the given date range. You can customize the stock symbols and date range based on your preferences.

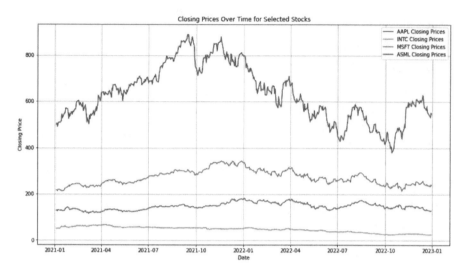

Figure 3-12. *Closing prices over time for selected stocks*

It is clear from Figure 3-12 that ASML has the highest closing rate even at it is lowest rates around October 2022, while Intel (INTC) has the lowest closing rate from January 2021 to January 2023.

Bar Chart

The bar chart is for comparing the values side to side whether vertical or horizontal. For this example, we shall use the crime data in NYC to compare the crime rate in the different NYC boroughs in 2018. You can download the 2018 NYPD Complaint Data from this link: `https://www.kaggle.com/datasets/mihalw28/nypd-complaint-data-current-ytd-july-2018/data?select=NYPD_Complaint_Data_Current_YTD.csv`.

For downloading the most recent data, it is available through the NYC Open Data Project website. You will find the data file in the Datasets Folder, or you can download the dataset from this link: `https://data.cityofnewyork.us/Public-Safety/NYPD-Complaint-Data-Current-Year-To-Date-/5uac-w243/data`.

To download the dataset of the current year from the above link, the steps are similar to the ones in Figure 3-13 by clicking on Export then CSV.

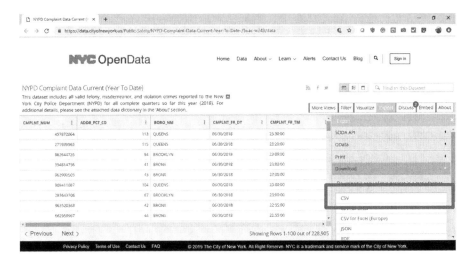

Figure 3-13. *NYPD Compliant Data—NYC OpenData*

The dataset contains a massive amount of information. Much investigation and analysis can be done over this data. For example: what is the highest crime type in every borough? When is the most probable time when crimes happen? Which day of the week where the highest number of crimes occur?

For this example, we are interested solely in answering: Which borough has the highest crime rate, and which has the least?

To answer this question, we shall need the following three columns: Borough, Longitude, and Latitude. Hence, we shall drop all the other columns. Then, we shall count the number of crimes in every borough to represent it with a bar chart.

After cleaning the data and preparing it, we have the following dataset as in Table 3-2.

Table 3-2. Number of Crimes Per Borough

Borough	Count of Crimes	Latitude	Longitude
BRONX	50153	40.82452	-73.8978
BROOKLYN	67489	40.82166	-73.9189
MANHATTAN	56691	40.71332	-73.9829
QUEENS	44137	40.74701	-73.7936
STATEN ISLAND	10285	40.63204	-74.1222

To load the data, we write Listing 3-5-1. In this example, we insert the values in the code instead of reading external data to practice other ways to read data in Python.

Listing 3-5-1. Creating a dataset for number of crimes per borough

```
# Data
Borough = ['BRONX', 'BROOKLYN', 'MANHATTAN', 'QUEENS', 'STATEN
ISLAND']
crimes = [50153, 67489, 56691, 44137, 10285]

# Create a bar chart
plt.figure(figsize=(10, 6))
plt.bar(Borough, crimes, color='skyblue')

# Add labels and title
plt.xlabel('Borough')
plt.ylabel('Number of Crimes')
plt.title('Number of Crimes in Different Boroughs')

# Display the bar chart
plt.show()
```

In Listing 3-5-1, we insert the names of the boroughs and the total number of crimes in each one in two lists, Borough and crimes. Then, we create the bar chart using the plt.bar() method and set the two variables and the color. Finally, as always, we set the labels of the x and y axes and the plot title and then show the plot.

The output of Listing 3-5-1 is Figure 3-14.

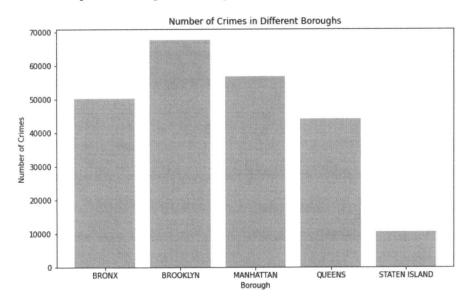

Figure 3-14. *Bar chart showing the number of crimes in different NYC boroughs*

How to Sort a Bar Chart?

To avoid any confusion, it is better to sort the graph in ascending or descending order. In this example, we shall learn a way to sort the previous bar chart of the crimes in NYC boroughs. There are many ways to sort the bar chart. In Listing 3-5-2, we see one of these ways.

Listing 3-5-2. Sort the vertical bar chart ascendingly

```
# Data
Borough = ['BRONX', 'BROOKLYN', 'MANHATTAN', 'QUEENS', 'STATEN
ISLAND']
crimes = [50153, 67489, 56691, 44137, 10285]

# Sort data by number of crimes
sorted_data = sorted(zip(Borough, crimes), key=lambda x: x[1])

# Extract sorted Borough and crimes after sorting
sorted_Borough, sorted_crimes = zip(*sorted_data)

# Create a bar chart
plt.figure(figsize=(10, 6))
plt.bar(sorted_Borough, sorted_crimes, color='skyblue')

# Add labels and title
plt.xlabel('Borough')
plt.ylabel('Number of Crimes')
plt.title('Number of Crimes in Different Boroughs (Sorted)')
# Display the bar chart
plt.show()
```

In Listing 3-5-2, we use zip(Borough, crimes) function to combine the two lists into pairs of boroughs and the count of crimes. The sorted() function sorts these pairs based on the second element of each pair, number of crimes, using a lambda function as the sorting key.

After sorting, the zip(*sorted_data) function is used to unzip the sorted data into two separate lists sorted_Borough and sorted_crimes. Then, as usual, we use Matplotlib to create a bar chart. The plt. figure(figsize=(10, 6)) sets the figure size. The plt.bar() function is then used to create a bar chart, where sorted_Borough is on the x axis, `sorted_crimes` variable is on the y axis, and the bars are colored sky blue. The `plt.xlabel`, `plt.ylabel`, and `plt.title` functions add labels to the x axis, y axis, and the title of the chart, respectively. Finally, `plt.show()` is used to display the created bar chart. The output of Listing 3-5-2 is the ascending sorted graph in Figure 3-15.

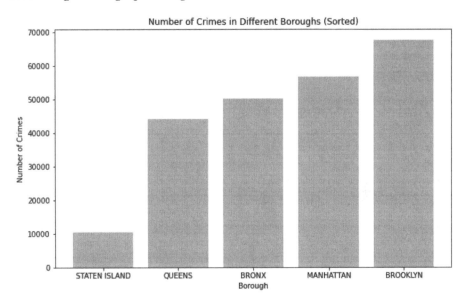

Figure 3-15. *Ascending sorted bar chart*

To reverse the output to get a descending vertical bar chart, set the reverse option to `True` in the `sorted_data = sorted(zip(Borough, crimes), key=lambda x: x[1], reverse=True)`, as in Listing 3-5-3.

Listing 3-5-3. Sort the bar chart in descending order

```
# Data
Borough = ['BRONX', 'BROOKLYN', 'MANHATTAN', 'QUEENS', 'STATEN
ISLAND']
crimes = [50153, 67489, 56691, 44137, 10285]

# Sort data by number of crimes in descending order
sorted_data = sorted(zip(Borough, crimes), key=lambda x: x[1],
reverse=True)

# Extract sorted Borough and crimes after sorting
sorted_Borough, sorted_crimes = zip(*sorted_data)

# Create a bar chart
plt.figure(figsize=(10, 6))
plt.bar(sorted_Borough, sorted_crimes, color='skyblue')

# Add labels and title
plt.xlabel('Borough')
plt.ylabel('Number of Crimes')
plt.title('Number of Crimes in Different Boroughs (Sorted
Descending)')

# Display the bar chart
plt.show()
```

The output of Listing 3-5-3 is in Figure 3-16.

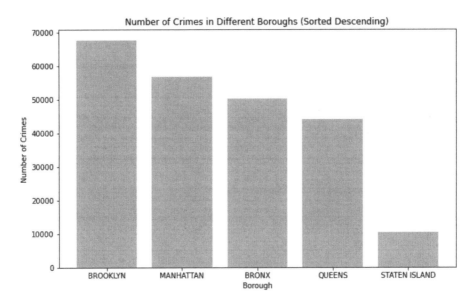

Figure 3-16. *Set reverse to true to reverse the bar chart in descending order*

To change the bar chart orientation from vertical to horizonal, we use `plt.barh()` function to plot the bar chart, as in Listing 3-5-4.

Listing 3-5-4. Sorted horizontal bar chart instead of a vertical one

```
# Data
Borough = ['BRONX', 'BROOKLYN', 'MANHATTAN', 'QUEENS', 'STATEN
ISLAND']
crimes = [50153, 67489, 56691, 44137, 10285]

# Sort data by number of crimes in descending order
sorted_data = sorted(zip(Borough, crimes), key=lambda x: x[1],
reverse=True)

# Extract sorted Borough and crimes after sorting
sorted_Borough, sorted_crimes = zip(*sorted_data)
```

```
# Create a horizontal bar chart
plt.figure(figsize=(10, 6))
plt.barh(sorted_Borough, sorted_crimes, color='skyblue')

# Add labels and title
plt.xlabel('Number of Crimes')
plt.ylabel('Borough')
plt.title('Number of Crimes in Different Boroughs (Sorted
Descending)')

# Display the horizontal bar chart
plt.show()
```

The output of Listing 3-5-4 is Figure 3-17.

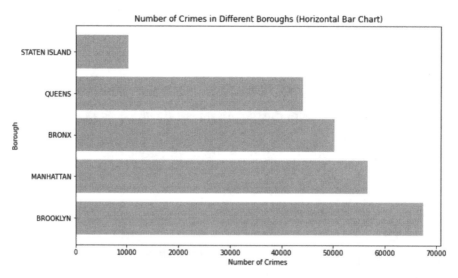

Figure 3-17.   *Sorted horizontal bar chart instead of a vertical one*

Create a Histogram Using a Bar Chart

A bar chart can be used to create histograms as well. In many other statics programming languages, as in SAS, the histograms can be drawn for a string data type. It must be numeric only. In Python, there is no such restriction. In this example, you will learn how to create the chart of the presidential candidates and their winning percentages in Maine in 2012, which was previously mentioned in Chapter 1. The results in Maine are as in Table 3-3.

Table 3-3. *Presidential Candidates in 2012 and Their Winning Percentages in Maine*

Candidates	Winning Percentages
Obama	56.27
Romney	40.98
Johnson	1.31
Stein	1.14
Paul	0.29
Anderson	0.01
Reed	0

Again, instead of loading the data from an external file, we shall insert the values in the code as in Listing 3-6.

Listing 3-6. Winning percentages of presidential candidates in Maine in 2012

```
# Data
candidates = ['Obama', 'Romney', 'Johnson', 'Stein', 'Paul',
'Anderson', 'Reed']
winning_percentages = [56.27, 40.98, 1.31, 1.14, 0.29, 0.01, 0]
```

```
# Define colors for each candidate
colors = ['blue', 'red', 'silver', 'green', 'grey',
'grey', 'grey']

# Create a bar chart
plt.figure(figsize=(10, 6))
bars = plt.bar(candidates, winning_percentages, color=colors)

# Add labels on top of each bar
for bar, label in zip(bars, winning_percentages):
    plt.text(bar.get_x() + bar.get_width() / 2 - 0.05,
    bar.get_height() + 0.2, f'{label:.2f}%', ha='center',
    va='bottom', color='black', fontweight='bold')

# Add labels and title
plt.xlabel('Candidates')
plt.ylabel('Winning Percentages (%)')
plt.title('Winning Percentages of Presidential Candidates')

# Display the bar chart
plt.show()
```

The added value of this bar chart in Listing 3-6 is coloring the bars with the corresponding party color. Again, we read the data from the two lists, candidates and winning_percentages. Then, we define a list of colors that we want to use in our chart. The plt.figure(figsize=(10, 6)) sets the figure size. The plt.bar function is then used to create a vertical bar chart, where the candidates column is on the x axis, the winning_percentages variable is on the y axis, and the bars are colored according to the specified colors' list that we pre-assigned. Then, we loop to iterate through each bar and its corresponding winning percentage.

The plt.text function is then used to add a text label above each bar, displaying the winning percentage with two decimal places. The labels are centered using ha='center' and aligned at the bottom using va='bottom'.

The labels are also bold and black. Then, at the end, as always, set the labels of the x and y axes and the plot title. The output is Figure 3-18.

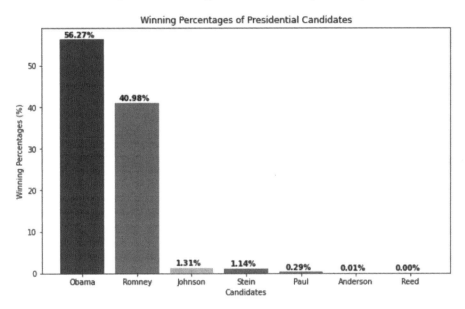

Figure 3-18. *The bars with the corresponding party color*

Bubble Chart

The bubble chart is useful in displaying three variables of data at a time. In this example, we will create a bubble chart to display the height, weight, and the age of a class.

For this example, I generated random data for the class, as shown in Listing 3-7-1, to function as a third data source in this chapter.

Listing 3-7-1. Randomize the class data and draw a bubble

```
import pandas as pd
import numpy as np
import matplotlib.pyplot as plt
```

```python
# Generate a dataset for 20 students
np.random.seed(42)  # Set seed for reproducibility

students_data = {
    'Name': [f'Student{i}' for i in range(1, 21)],
    'Height': np.random.uniform(150, 190, 20),  # Heights in
                                         centimeters
    'Weight': np.random.uniform(45, 90, 20),   # Weights in
                                         kilograms
    'Age': np.random.randint(18, 25, 20)        # Ages in years
}

students_df = pd.DataFrame(students_data)

# Create a bubble chart
plt.figure(figsize=(12, 8))

# Adjust the scaling factor for bubble sizes
scaling_factor = 500

scatter = plt.scatter(students_df['Height'], students_
df['Weight'], s=students_df['Age'] * scaling_factor,
c=students_df['Age'], cmap='viridis', alpha=0.7)

# Annotate each point with student name
for i, name in enumerate(students_df['Name']):
    plt.text(students_df['Height'][i], students_df['Weight']
    [i], name, ha='center', va='center', fontsize=8,
    color='black', alpha=0.7)

# Add labels and title
plt.xlabel('Height (cm)')
plt.ylabel('Weight (kg)')
plt.title('Bubble Chart: Height vs Weight with Age as Bubble
Size and Student Names')
```

```
# Add colorbar
cbar = plt.colorbar(scatter)
cbar.set_label('Age (years)')

# Display the bubble chart
plt.show()
```

Listing 3-7-1 starts with setting the random seed for reproducibility using np.random.seed(42). Making the seed fixed in the randomization means that every time we recompile and re-run the code, the same random values will be generated. But if we do not specify a seed, every time we re-run the cell, new random values will be generated. Then, use np.random.uniform function to generate random data for 20 students. The dataset includes columns for student names, heights, weights, and ages. Then, create a bubble chart using Matplotlib. As always, the plt. figure(figsize) sets the figure size.

Now, the surprise is using plt.scatter function again. Yes, we use it for drawing scatter and bubble charts. In the scatterplot, the x axis represents heights, the y axis represents weights, the bubble sizes are proportional to ages scaled by a factor of 500, and the colors represent ages. The cmap='viridis' specifies the color map, and alpha=0.7 sets the transparency of the bubbles.

Then use the plt.text function to add annotations for each point on the chart, displaying the student names at the corresponding positions. The code adds labels to the x axis, y axis, and a title to the chart.

A colorbar legend is added to the chart to represent the ages of students. The cbar.set_label function sets the label for the colorbar. Finally, plt.show() is used to display the created bubble chart. The output is Figure 3-19.

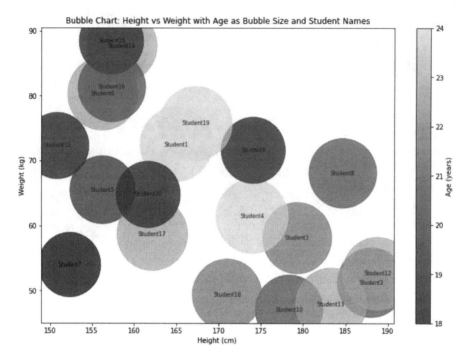

Figure 3-19. *The initial bubble chart*

As you see in Figure 3-19, the bubble diameters do not have a substantial difference. The ages are between 18 and 24 years old. Therefore, I decided to save one of the generated random datasets in an Excel file and manually changed some of the age, weight, and height values to make a significant difference. The age ranges between 5 and 24 years old, as in class.xlsx dataset in the Datasets Folder, as in Listing 3-7-2. The only difference between Listings 3-7-1 and 3-7-2 is replacing the randomization part by creating a DataFrame, students_df, by reading the file using pd.read_excel function.

Listing 3-7-2. Reading the class dataset instead of the randomization

```python
import pandas as pd
import numpy as np
import matplotlib.pyplot as plt

# Generate a dataset for 20 students
np.random.seed(42)  # Set seed for reproducibility

students_df = pd.read_excel('../Datasets/Chapter 3/class.xlsx')
print(df.head())')

# Create a bubble chart
plt.figure(figsize=(12, 8))

# Adjust the scaling factor for bubble sizes
scaling_factor = 500

scatter = plt.scatter(students_df['Height'], students_
df['Weight'], s=students_df['Age'] * scaling_factor,
c=students_df['Age'], cmap='viridis', alpha=0.7)

# Annotate each point with student name
for i, name in enumerate(students_df['Name']):
    plt.text(students_df['Height'][i], students_df['Weight']
    [i], name, ha='center', va='center', fontsize=8,
    color='black', alpha=0.7)

# Add labels and title
plt.xlabel('Height (cm)')
plt.ylabel('Weight (kg)')
plt.title('Bubble Chart: Height vs Weight with Age as Bubble
Size and Student Names')
```

```
# Add colorbar
cbar = plt.colorbar(scatter)
cbar.set_label('Age (years)')

# Display the bubble chart
plt.show()
```

The output of Listing 3-7-2 is Figure 3-20. Now, the chart clearly displays the class outliers. For example, Student 5 in a purple bubble is the smallest student in the class vs. the biggest student, Student 4, in a yellow one.

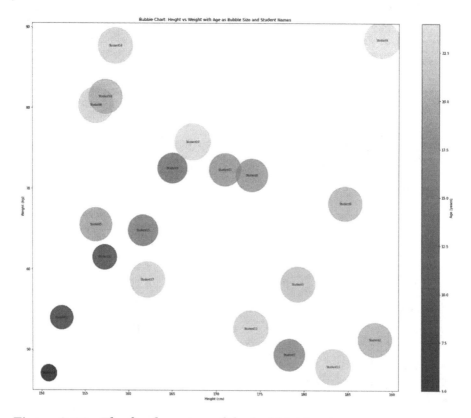

Figure 3-20. *The final version of the bubble chart*

Summary

This chapter digs deeper into data visualization. It starts with the most essential plots, such as scatterplots and histograms. Moreover, it shows how to concatenate plots over each other by changing the colors and patterns to ease the comparisons of the findings. Data visualization in Python is one of the most powerful features and has plenty of options to customize plots.

CHAPTER 4

Statistical Analysis and Linear Models

In Chapter 3, we explored graph and data visualization. In Chapter 4, we will explore statistics and linear models.

Statistical Analysis

After reading datasets, there are usual statistical analysis tasks that should be performed to understand the data and to be included in the reports. In this chapter, we will discuss frequency tables, summary statistics, correlation analysis, and T-tests.

Frequency Tables

The frequency tables task is usually from the first steps when exploring any dataset. It generates frequency tables from your data showing the unique levels of each variable. Also, the table can help you identify the outliers.

For this example, we explore the HEART.csv in the Datasets Folder to see how many people have high cholesterol. To do that, we focus on CHOL_STATUS, which is a string variable. The code is in Listing 4-1 in Chapter 4.ipynb in the Example Code Folder.

© Engy Fouda 2024
E. Fouda, *Learn Data Science Using Python*, https://doi.org/10.1007/979-8-8688-0935-4_4

In Listing 4-1, we learn how to generate a frequency table and plot the frequency distribution in a bar chart.

Listing 4-1. Frequency tables

```
import pandas as pd
import matplotlib.pyplot as plt

heart_df = pd.read_csv('../Datasets/Chapter 4/HEART.csv')
frequency_table = heart_df['Chol_Status'].value_counts().reset_
index().rename(columns={'index': 'Chol_Status', 'Chol_Status':
'Frequency'})

# Sort the table by frequency
frequency_table = frequency_table.sort_values(by='Frequency',
ascending=False)

# Print the frequency table
print(frequency_table)

# Plot the frequency
plt.figure(figsize=(8, 6))
plt.bar(frequency_table['Chol_Status'], frequency_
table['Frequency'], color='skyblue')
plt.xlabel('Chol_Status')
plt.ylabel('Frequency')
plt.title('Frequency Plot of Chol_Status')
plt.show()
```

After importing the required libraries and reading the csv file in a Panda DataFrame, heart_df, we calculate the frequency table. This code line, frequency_table = heart_df['Chol_Status'].value_counts().reset_index().rename(columns={'index': 'Chol_Status', 'Chol_Status': 'Frequency'}), performs the following:

- `heart_df['Chol_Status'].value_counts()`: This calculates the frequency of unique values in the `'Chol_Status'` column of the DataFrame `heart_df`.

- `.reset_index()`: This resets the index of the resulting series, converting it into a DataFrame.

- `.rename(columns={'index': 'Chol_Status', 'Chol_Status': 'Frequency'})`: This renames the columns of the DataFrame to `'Chol_Status'` and `'Frequency'` for clarity.

Then, we sort the frequency table DataFrame by the `Frequency` column in descending order, ensuring that the categories with the highest frequencies are displayed first. This is done by this instruction: `frequency_table = frequency_table.sort_values(by='Frequency', ascending=False)`.

Then, we print the frequency table DataFrame to the console, showing the frequencies of each unique value in the `Chol_Status` column. The output of this code is in the table in Figure 4-1.

```
  Chol_Status  Frequency
0   Borderline       1861
1         High       1791
2    Desirable       1405
```

Figure 4-1. *Frequency table*

For plotting the frequency plot, as mentioned before, we use `plt.figure()` method to create a new plot and do not have the plots drawn over each other. Also, we use it to set the plot dimensions. Then, create a bar chart of `Chol_Status` series as the x axis and the frequency as the y axis and add their labels. The last two steps are for setting the plot title and showing it. The output of this part is in Figure 4-2.

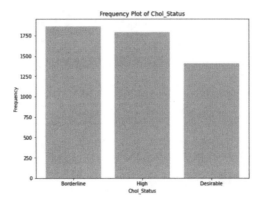

Figure 4-2. *Frequency bar chart*

Figure 4-2 shows the distribution of Chol_Status in descending order, where Borderline has the highest percentage in this dataset, followed by High, and the lowest frequency is Desirable.

Summary Statistics

The summary statistics task provides descriptive statistics for variables across all observations and within groups of observations. You can also summarize your data in a graphical display, such as a histogram or a box plot.

For this example, we review Maine's past elections to see if there are any predictable outcomes (patterns). The dataset is in the Datasets folder and has the name Maine_Past_Elections.xlsx. The contents of the Excel file are as shown in Table 4-1.

Table 4-1. *Maine_Past_Elections.xlsx*

Year	D	%	R	%
2016	**357,735**	**47.80%**	335,593	44.90%
2012	**401,306**	**56.27%**	292,276	40.98%
2008	**421,923**	**57.71%**	295,273	40.38%
2004	**396,842**	**53.57%**	330,201	44.58%
2000	**319,951**	**49.10%**	286,616	44.00%
1996	**312,788**	**51.60%**	186,378	30.80%
1992	**263,420**	**38.80%**	206,820	30.40%
1988	243,569	43.90%	**307,131**	**55.30%**
1984	214,515	38.80%	**336,500**	**60.80%**
1980	220,974	42.30%	**238,522**	**45.60%**
1976	232,279	48.07%	**236,320**	**48.91%**
1972	160,584	38.50%	**256,458**	**61.50%**
1968	**217,312**	**55.30%**	169,254	43.10%
1964	**62,264**	**68.84%**	118,701	31.16%
1960	**181,159**	**42.95%**	240,608	57.05%

Let us read the data and compute the summary statistics for both columns of Democrats and Republicans, D and R variables in the dataset, as in Listing 4-2.

Listing 4-2. Summary statistics

```
import pandas as pd
import numpy as np
import matplotlib.pyplot as plt
```

```
past_elections_df = pd.read_excel('../Datasets/Chapter 4/Maine_
Past_Elections.xlsx')
print(past_elections_df['D'].describe())
print(past_elections_df['R'].describe())
print(past_elections_df.isnull().sum())
```

After importing the libraries and loading the dataset, we compute and print descriptive statistics for columns D and R in the DataFrame past_elections_df. The .describe() method generates summary statistics such as count, mean, standard deviation, minimum, and maximum for the specified column.

This .isnull() method checks for missing values in the DataFrame past_elections_df, which returns a DataFrame of Boolean values indicating the presence of missing values. The .sum() method then calculates the sum of missing values for each column and prints the result. The output is Figure 4-3. As shown, there are no null values in the dataset.

```
count        15.000000
mean     267108.066667
std      100623.204581
min       62264.000000
25%      215913.500000
50%      243569.000000
75%      338843.000000
max      421923.000000
Name: D, dtype: float64
count        15.000000
mean     255776.733333
std       65083.494267
min      118701.000000
25%      221570.000000
50%      256458.000000
75%      301202.000000
max      336500.000000
Name: R, dtype: float64
Year    0
D       0
%       0
R       0
%.1     0
dtype: int64
```

Figure 4-3. *Summary statistics of Democrats and Republicans in Maine*

The difference between the Democrats' and Republicans' averages is not remarkable. Therefore, Maine is "Lean Democrat," as Democrats won 7 elections out of 15 (from the year 1960 to 2016). Since 1992, the Democratic Party has always won the elections in Maine.

Correlation Analysis

Before explaining the correlation concepts, we want to set the difference between the causation and the correlation. Causation is when one thing causes another thing to happen, while correlation is when two or more things appear to be related. Correlations do not always mean causation.

For example, travel causes buying suitcases and tickets, but suitcase sales and tickets are correlated, as shown in Figure 4-4.

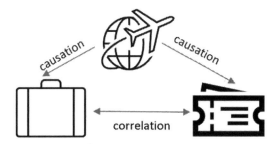

Figure 4-4. *Causation vs. correlation*

Correlation serves as a statistical method to depict the connection between numerical variables. This connection is articulated through the computation of correlation coefficients, which range from -1 to 1. The correlation analysis task furnishes both graphs and statistics to explore the associations among variables.

Let us inspect a couple of examples. The first one is finding if there is a correlation between the length and weight of cars and their horsepower. The cars' dataset is in the Datasets Folder.

The code is in Listing 4-3 in Chapter 4.ipynb in the Example Code Folder. There are multiple ways to compute correlation in Python. In this example, we use the corr() method in Pandas to compute the correlation matrix between the specified variables.

In Listing 4-3, we import the required libraries, load the dataset, and explore it. To explore the data more, we plot the multiple pairwise bivariate distributions of the variables.

Listing 4-3. Import the required libraries, load the dataset, and explore it

```
import pandas as pd
import numpy as np
import matplotlib.pyplot as plt

cars_df = pd.read_excel('../Datasets/Chapter 4/CARS.xlsx')
print(cars_df.head())
print(cars_df.describe())

# Selecting variables
variables = cars_df[['Weight', 'Length', 'Horsepower']]

# Calculate the correlation matrix
correlation_matrix = variables[['Weight', 'Length']].corrwith(v
ariables['Horsepower'])

print("Pearson correlation coefficients:")
print(correlation_matrix)

plt.figure(figsize=(12, 6))
plt.scatter(cars_df['Weight'], cars_df['Horsepower'])
plt.xlabel('Weight')
plt.ylabel('Horsepower')
```

```
plt.figure(figsize=(12, 6))
plt.scatter(cars_df['Length'], cars_df['Horsepower'])
plt.xlabel('Length')
plt.ylabel('Horsepower')
```

In Listing 4-3, we start by importing the Pandas library, which is used for data manipulation. We load the dataset CARS.xlsx into a Pandas DataFrame using the pd.read_excel() function. Then, we extract the columns Weight, Length, and Horsepower from the dataset and store them in a new DataFrame called variables.

We use the corrwith() method to compute the Pearson correlation coefficients between the pairs of variables Weight and Horsepower and Length and Horsepower. This method calculates the correlation between two DataFrames or two Series. In this case, we correlate the Weight and Length columns in the variables DataFrame with the Horsepower column. The output in Figure 4-5 shows that the horsepower is more correlated with weight than length.

```
Pearson correlation coefficients:
Weight     0.630796
Length     0.381554
dtype: float64
```

Figure 4-5. *Pearson correlation coefficients*

To confirm these correlations, let us draw the scatterplots, as in Figure 4-6.

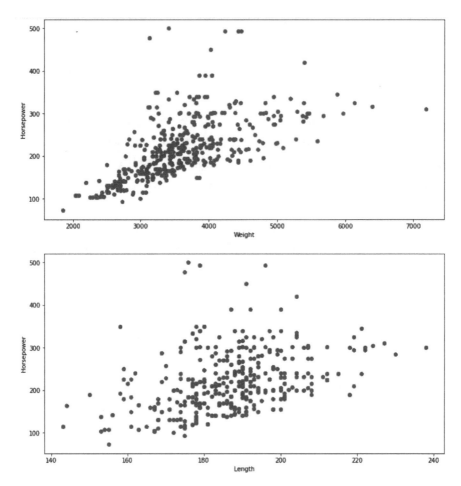

Figure 4-6. *Correlation between horsepower and weight and length*

Now, let us explore how correlation can be applied to the presidential elections in Maine. The inquiry at hand is: Does the gross domestic product (GDP) have an impact on voter turnout in the state of Maine?

The dataset named correlation.xlsx is available in the Datasets directory. However, constructing this dataset posed a challenge. Different media outlets and government sources presented varying figures for Maine's voter turnout. For this analysis, I opted to utilize the most frequently cited numbers from 2016. It is important to note that further validation is warranted if these figures are to be utilized beyond the scope of this exercise. Below is a sample of the sources I consulted:

- https://en.wikipedia.org/wiki/United_States_presidential_election_in_Maine,_2000

- http://www.270towin.com/states/Maine

- https://www.pressherald.com/2016/11/08/mainers-head-to-polls-in-historic-election/

- https://www.mainepublic.org/politics/2016-11-08/maine-voter-turnout-extremely-heavy

- https://www.csmonitor.com/USA/Elections/2012/1106/Voter-turnout-the-6-states-that-rank-highest-and-why/Maine

- https://www.census.gov/history/pdf/2008presidential_election-32018.pdf

- https://bipartisanpolicy.org/report/2012-voter-turnout/

- https://www.census.gov/content/dam/Census/library/publications/2008/demo/p20-557.pdf

- https://www.census.gov/content/dam/Census/library/publications/2006/demo/p20-556.pdf

- https://www.census.gov/content/dam/Census/library/publications/2002/demo/p20-542.pdf

- https://www.deptofnumbers.com/gdp/maine/

Listing 4-4. Is the GDP correlated with Maine voter turnout?

```
import pandas as pd
import numpy as np
import matplotlib.pyplot as plt

correlation = pd.read_excel('../Datasets/Chapter 4/
correlation.xlsx')
print(correlation.head())
print(correlation.describe())

# Calculate the correlation matrix
correlation_coefficient = correlation['gdp'].corr(correlation
['voterturnout'])

#print("Pearson correlation coefficients:")
print("Pearson correlation coefficient between GDP and Voter
Turnout:", correlation_coefficient)

plt.figure(figsize=(12, 6))
plt.scatter(correlation['gdp'], correlation['voterturnout'])
plt.xlabel('gdp')
plt.ylabel('voterturnout')
```

Listing 4-4 begins by importing the necessary libraries. It then reads the dataset correlation.xlsx into a Pandas DataFrame named `correlation` using the `pd.read_excel()` function. As usual, we verify that the dataset has been read correctly by using the `print(correlation.head())` statement to display the first few rows of the dataset, giving an overview of its structure and contents. This helps in understanding the data's format and identifying any potential issues.

Similarly, `print(correlation.describe())` statement prints the summary statistics of the dataset, including count, mean, standard deviation, minimum, and maximum values for each numeric column. This gives insights into the distribution and range of values in the dataset.

The code calculates the Pearson correlation coefficient between the
gdp (Gross Domestic Product) and voterturnout columns using the
.corr() method. This coefficient measures the strength and direction of
the linear relationship between two variables. The result is stored in the
variable correlation_coefficient. The output of the above statements is
shown in Figure 4-7.

```
   year  voterturnout      gdp
0  2016          67.0   51.823
1  2012          67.2   50.106
2  2008          70.2   51.236
3  2004          72.0   51.351
4  2000          69.2   46.618
              year  voterturnout         gdp
count     5.000000      5.000000    5.000000
mean   2008.000000     69.120000   50.226800
std       6.324555      2.100476    2.113539
min    2000.000000     67.000000   46.618000
25%    2004.000000     67.200000   50.106000
50%    2008.000000     69.200000   51.236000
75%    2012.000000     70.200000   51.351000
max    2016.000000     72.000000   51.823000
Pearson correlation coefficient between GDP and Voter Turnout: 0.04994549957956506
```

Figure 4-7. *The .head(), .describe(), and .corr() methods' outputs*

As shown in Figure 4-7, the correlation coefficient is slightly less than
0.05, which indicates that there is almost no relation between the GDP and
the voter turnout in Maine.

In Listing 4-4, we visually explore the relationship between GDP and
voter turnout; a scatterplot is created using Matplotlib. The .scatter()
method plots the gdp on the x axis and the voterturnout on the y axis. The
plt.xlabel() and plt.ylabel() functions set labels for the x axis and
y axis, respectively, to provide context to the plot. The .title() method
sets a title for the plot. Finally, plt.show() displays the plot. On the plot,
you will find the points are all over the place indicating that there is no
relationship between Maine voters' turnout and the GDP, as shown in
Figure 4-8.

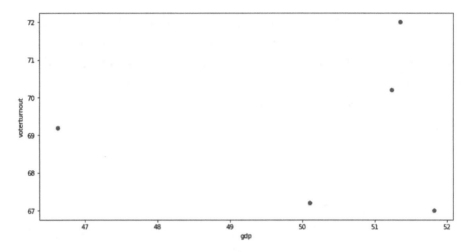

Figure 4-8. *Correlation between Maine voter turnout and GDP*

Hypothesis Testing

Hypothesis testing is a statistical method used to make inferences about a population parameter based on sample data. It involves comparing observed data to an expected or null hypothesis to determine whether any observed differences or relationships are statistically significant or simply due to random chance.

Examples of hypothesis testing for continuous data analysis are T-test and ANOVA, while chi-square test is for categorical data analysis.

T-Test

Frequently, we seek to ascertain whether there exists a genuine disparity in means between two distinct groups or if the perceived distinction is merely a result of random variation. The T-test serves this purpose by gauging the likelihood of disparity between two datasets. In other words, it assesses the mean values of two samples. It is worth noting that the T-test is employed for small sample sizes since they may not adhere to the normal distribution.

One-Sample T-Test

In this example, I gathered all the voting data from past presidential elections in Maine for candidates representing both major political parties across various years. This dataset served as the basis for predicting the election outcomes in Maine for both the 2016 and 2020 elections.

The dataset, MaineVotesDR.xlsx, comprises four columns: Election Year, Democratic Votes, Republican Votes, and Result. The Result column records a binary outcome: 0 signifies victory for the Republican candidate, while 1 indicates success for the Democratic candidate.

Our hypothesis is that the Republican candidate will be victorious in the 2020 presidential elections in Maine, while the alternative hypothesis suggests the victory of the Democratic candidate.

Null hypothesis (H0): result = 0Alternative hypothesis (Ha): result = 1

Now, let us load the dataset and explore how the T-test can shed light on the outcomes of the 2020 election, as shown in Listing 4-5.

Listing 4-5. One-sample T-test

```python
import pandas as pd
from scipy import stats

maine_votes_df = pd.read_excel('../Datasets/Chapter 4/
MaineVotesDR.xlsx')
# Test for normality
normality_test = stats. shapiro (maine_votes_df['Result'])
print("Normality Test:")
print(normality_test)

# t-test
t_stat, p_value = stats.ttest_1samp(maine_votes_df['Result'], 0)
print("\nT-test:")
print("T-statistic:", t_stat)
print("P-value:", p_value)
```

In Listing 4-5, we use `stats.shapiro()` to perform the normality test on the Result column of the DataFrame. It is appropriate for smaller sample sizes. The results will give a test statistic and a p-value to determine if the data deviates significantly from normality. If the dataset is larger than 20 rows, the `stats.normaltest()` function can be used. Then, we use `stats.ttest_1samp()` to perform the one-sample t-test on the Result column with the null hypothesis (h0) being 0. We print the results of both tests. The output is as in Figure 4-9.

```
Normality Test:
ShapiroResult(statistic=0.6303409337997437, pvalue=4.9036381824407727e-05)

T-test:
T-statistic: 4.58257569495584
P-value: 0.0004263757573432252
```

Figure 4-9. *T-test results: prediction for 2020 presidential elections in Maine*

The p-value = 0.0004. Since the p-value is less than 0.05, we can reject the null hypothesis. Hence, we can conclude that the Republican candidate will lose in Maine in the 2020 election. He did lose Maine.

Two-Sample T-Test

In this example, the question is: Is there a significant difference in height between the two genders?

The null hypothesis is that there is no difference in height. The alternate hypothesis is that there is a difference in height based on gender.

In this example, load the CLASS dataset, as in Listing 4-6-1.

Listing 4-6-1. Two-sample T-test

```
import pandas as pd
import numpy as np
from scipy import stats
```

```python
import seaborn as sns
import matplotlib.pyplot as plt
#https://www.kaggle.com/datasets/megasatish/class-data

class_df = pd.read_csv('../Datasets/Chapter 4/Class.csv')

# Test for normality by Gender
print("Normality Test by Gender:")
for Gender in class_df['Gender'].unique():
    stat, p_value = stats. shapiro (class_df[class_df['Gender']
    == Gender]['Height'])
    print(f"Gender: {Gender}, Statistic: {stat:.2f}, p-value:
    {p_value:.4f}")

# t-test by Gender
# comparing males to females
male_heights = class_df[class_df['Gender'] == 'M']['Height']
female_heights = class_df[class_df['Gender'] == 'F']['Height']
t_stat, p_value = stats.ttest_ind(male_heights, female_heights)
print(f"\nT-test between male and female Heights:")
print(f"T-statistic: {t_stat:.2f}, p-value: {p_value:.4f}")
```

In Listing 4-6-1, after importing the required libraries and loading the dataset, we test for the normality of heights within each gender group using the .shapiro() method as in the previous listing from scipy.stats. Then, we loop through unique gender values ("M" and "F") to perform the normality test on the heights for each gender. And we print the test statistic and p-value for each gender, as Figure 4-10.

```
Shapiro Normality Test by Gender:
Gender: M, Statistic: 0.95, p-value: 0.7249
Gender: F, Statistic: 0.93, p-value: 0.4932


T-test between male and female Heights:
T-statistic: 1.45, p-value: 0.1645
```

Figure 4-10. *Normality test by gender*

From Figure 4-10, we cannot reject the null hypothesis as there is not enough evidence for that; the p-value is 0.1645, which is larger than 0.05. We can conclude that there is no difference in height between the two genders. However, we must note that the sample is small. The CLASS dataset contains only 19 rows.

To understand the data distribution more, let us visualize the height of each gender using box and line plots. As in Listing 4-6-2.

Listing 4-6-2. Visualizing the height of each gender

```
# Visualizing the Height distribution by Gender
sns.boxplot(x='Gender', y='Height', data=class_df)
plt.title('Height Distribution by Gender')
plt.show()

# For t-test plots, let us use seaborn to show the distribution
of heights by Gender,
# highlighting the means and confidence intervals
sns.catplot(x='Gender', y='Height', kind="point", data=class_
df, join=False, capsize=.1)
plt.title('Point Plot Showing Height Means by Gender')
plt.show()
```

In Listing 4-6-2, we use the `sns.boxplot()` method to plot the height distribution of each gender. As in Figure 4-11, the box plot shows that there is a small difference in the medians of the two groups, almost negligible. Also, there is a small difference between the means of the two groups, as shown in the line plot.

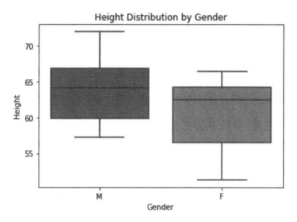

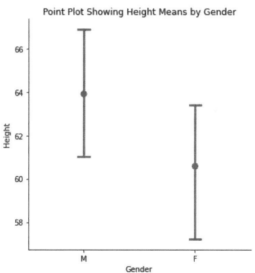

Figure 4-11. *Box plot and line plot for the height distribution by Gender*

Summary

This chapter explains the most crucial statistical concepts that will be needed for almost all analytics and data science reports. They are the frequency tables, summary statistics, correlation analysis, and T-tests. In the next chapter, we shall learn the operators and loops.

## CHAPTER 5

Data Preprocessing and Feature Engineering

This chapter shows some essential data cleaning and querying methods, such as comment statement, arithmetic, and comparison operators, and how to represent missing values and loops.

Comment Statement

It is crucial to document your code so you and your peers can understand the reasoning behind how the analysis is constructed. In general, if you do not document your code, you might look at it after a few months and not remember why you used a certain function or why you made a certain check. Hence, code documentation is crucial to follow the logic flow.

The syntax of the comment statement in Python can be done in two ways as follows:

```
#message
"""message"""
```

You can use the comment statement anywhere in the program to explain your code. The Python interpreter ignores text in comment statements during processing. It is always recommended that you start your programs by writing a comment at the top explaining what your program will do and then add as many comments as you need. To describe the logic flow, it is best to write a comment at every data or procedure step to describe what it does.

Arithmetic and Assignment Operators

Arithmetic operators indicate that an arithmetic calculation is performed, as shown in Table 5-1.

Table 5-1. *Arithmetic and Assignment Operators*

Operator	Meaning	Example	Result
+	Addition	2 + 3	5
+=		X = 5X += 15	X=20
-	Subtraction	5 - 3	2
-=		X = 5X -= 15	X=-10
/	Division	5 / 2	2.5
/=		X = 15 X /= 5	X=3.0
//	Quotient	5 // 2	2
%	Remainder	5 % 2	1
*	Multiplication	2 * 3	6
*=		X = 15 X *= 5	X=75
**	Exponentiation	2 ** 3	8

If a missing value is an operand for an arithmetic operator, the result is a missing value. To avoid propagating missing values, we discuss some of the methods of how to represent and handle the missing values in your dataset later in this chapter and in the next one.

Python Operator Precedence Rule: PEMDAS

- P: Parentheses

- E: Exponentiation

- M: Multiplication

- D: Division

- A: Addition

- S: Subtraction

You can remember it with the popular order of operators' statement: "Please Excuse My Dear Aunt Sally."

Comparison Operators

Comparison operators set up a comparison, operation, or calculation with two variables, constants, or expressions. If the comparison is true, the result is True. If the comparison is false, the result is False. Comparison operators can be expressed as in Table 5-2.

Table 5-2. *Comparison Operators*

Operator	Meaning	Example	Result
==	Equality	print(2 == 3)	False
!=	Inequality	print(2 != 3)	True
<	Less than	print(2 < 3)	True
<=	Less than or equal	print(2 <= 3)	True
>	Greater than	print(2 > 3)	False
>=	Greater than or equal	print(2>=3)	False
in	Membership	x=2 print(x in (2,3,5))	True

Logical, Membership, and Identity Operators

Logical operators are used to show the relationship between two conditions or expressions, determining whether they are both true, at least one is true, or the condition should be reversed. Table 5-3 shows the three types of operators: logical, membership, and identity ones.

Membership operators are used to check if a value is part of a collection or sequence, like a list or string.

Identity operators are used to determine if two variables are actually the same object in memory, not just equivalent in value.

Table 5-3. *Logical, Membership, and Identity Operators*

Operator	Meaning	Example	Result
and	Returns True if both statements are true	x=5 x < 5 and x < 10	False
or	Returns True if one of the statements is true	x=5 x < 5 or x < 4	False
not	Reverses the result, returns False if the result is true	x=5 not(x < 5 and x < 10)	True
in	Returns True if a sequence with the specified value is present in the object	y=['bus','car','walk'] x='car' x in y	True
not in	Returns True if a sequence with the specified value is not present in the object	y=['bus','car','walk'] x='car' x not in y	False
is	Returns True if both variables are the same object	x=['bus','car','walk'] y=['bus','car','walk'] print(x is y) print(x==y)	False True
is not	Returns True if both variables are not the same object	x=['bus','car','walk'] y=['bus','car','walk'] x is not y	True

The order of operations in Python is as follows: parentheses, exponentiation, unary plus, unary minus, multiplication, division, floor division, and modulus, addition and subtraction, comparisons, identity, and membership operators, logical NOT, AND, OR.

How to Represent Missing Values in Raw Data

These are the values that represent missing values in raw data, so that the Python interpreter reads and stores the value appropriately:

```
null,  , NA, nan, NaN
```

The Python interpreter will convert the above values to NaN and handles it as a missing value. Please pay attention to the case, as the following values are not interpreted as missing: `None, none, na, Null`.

Let us test these values and see with an example how the Python interpreter computes them. In Listing 5-1, we test all the mentioned values above and some of the missing values' representations of other statistical programming languages. For example, "." represents the missing values in SAS.

Listing 5-1. Try different values to represent missing values, and see which ones the Python interpreter is able to read correctly

```python
#import the libraries
import pandas as pd
import numpy as np
import matplotlib.pyplot as plt

#import the dataset
students_df = pd.read_excel('../Datasets/Chapter 5/missing.xlsx')
#verify that the dataset has been read correctly
print(students_df)
#generate the summary statistics
students_df['Height'].describe()

#create a new empty dataframe that has the Height and Missing
variables
```

```
new=pd.DataFrame(columns = ['Height', 'Missing'])
#Height is equal the Haight column of the dataset
new['Height'] = students_df['Height']
#Missing equals the Boolean output of checking of each value is
missing or #not
new['Missing']=students_df['Height'].isnull()
#print the new dataframe after filling it in
print(new)
```

In Listing 5-1, we start by importing the necessary libraries. Then, import the dataset missing.xlsx. You can access it in the Datasets Folder. The whole code is in Listing 5-1 in Chapter 5.ipynb in the Example Code Folder. Figure 5-1 shows the values of the dataset, where the light gray background is for the correct missing values representations and the darker shade is for the incorrect ones.

No	Name	Height	Weight	Age
0	Student1	164.9816	72.53338	16
1	Student2		51.27722	21
2	Student3	null	58.14651	22
3	Student4	188.7964	88.45344	24
4	Student5	None	65.52315	20
5	Student6	156.2398	80.33292	23
6	Student7	none	49.39525	18
7	Student8	184.647	68.14055	21
8	Student9	na	71.65866	19
9	Student10	nan	53.98532	10
10	Student11	170.9903	72.33952	19
11	Student12	Null	52.67359	23
12	Student13	183.2977	47.92732	23
13	Student14	NA	87.69985	23
14	Student15	161.6492	64.80686	17
15	Student16	NaN	81.37788	21
16	Student17	162.1697	58.70762	23
17	Student18	.	47.09027	5
18	Student19	167.2778	75.79049	24
19	Student20	?	61.48628	11

Figure 5-1. *Missing values dataset*

Continuing Listing 5-1, after we import the above dataset, we print it to see how the values are imported. You will find that all the correct representations will be read as NaN. We use the describe() method to compute the summary statistics of the variable. In such case because of the many incorrect values, the Python interpreter cannot compute the statistics report.

To check each value side by side to the isnull() method output, we create a new DataFrame called new. It has two variables, Height and Missing. This series equals the Height column in the students_df DataFrame. Missing equals the output of the isnull() method. At the end, we print the DataFrame new to see the values side by side. The output is in Figure 5-2.

```
        Height  Missing
0    164.981605    False
1           NaN     True
2           NaN     True
3    188.796394    False
4          None    False
5    156.239781    False
6          none    False
7    184.647046    False
8            na    False
9           NaN     True
10   170.990257    False
11         Null    False
12   183.297706    False
13          NaN     True
14   161.649166    False
15          NaN     True
16    162.16969    False
17             .   False
18   167.277801    False
19            ?    False
```

Figure 5-2. *Checking the values side by side*

From Figure 5-2, we verified that Python only accepts `null, , NA,`
`nan, NaN` to represent the missing values in raw data.

Loops

There are two types of loops in Python for iteration: while and for loops.

- While Loop: You use it if you do not know the number of iterations and rely on a condition to be false to exit the loop.

- For Loop: You use it if you know the number of iterations, or loop over a certain container variable, or within a specific range of numbers.

Let us explore some examples for both.

While Loop

For this example of the while loop, we keep looping asking the user to input numbers, and the program will output if the number is even or odd. Again, the most important thing to think about when using the while loop is the exiting or breaking condition, in other words, when the loop will end; otherwise, it is an infinite loop and will generate an error. For this example, the exit condition is when the user enters zero. The program will display an exit message and branches.

The loops in Python have a unique feature that is not in other languages, which is using `else` with the while and for loops. In this example, we use the `while-else` to display the exit message, as in Listing 5-2.

Listing 5-2. While loop example

```
#input the first number from the user before the loop
input_no = int(input('Enter your number (0 to exit): '))
#check the input value if it equals zero or not; if not, the
while loop is #entered
while input_no != 0:
    #check the remainder after dividing by 2 if it is or
    zero or not
    if input_no%2==0:
        #if the remainder is zero, the input number is even
        print("your input is even")
    else:
        print("your input is odd")
    #ask the user to enter a new number
    input_no = int(input('Enter your number (0 to exit): '))
else:
    #it keeps looping till the user enters zero. It prints this
    message and      #exits
    print("Bye and Have a great day!")
```

In Listing 5-2, we start with asking the user to enter the first value. The program checks in the while condition if it equals zero or not. If the input is not equal to zero, it enters the loop and checks using the if-condition the modulus of the number. If the remainder of the modulus equals zero, then it is an even number; otherwise, it is odd.

If the input number equals zero, it goes to the else of the while loop. The program prints the message and exits. Figure 5-3 shows a sample output of the listing.

```
Enter your number (0 to exit): 2
your input is even
Enter your number (0 to exit): 15
your input is odd
Enter your number (0 to exit): 0
Bye and Have a great day!
```

Figure 5-3. *While loop example output*

For Loop

The for loop is more common than the while loop and is used to loop over containers, DataFrames, lines in a file, and more. In this section, we have some examples to show the usage of the for loop. The first example is the most standard looping example of printing a sequence of numbers and printing their sum. The second one is for looping over the letters of a word. The third one is looping over a list, as one of the containers' datatype examples. The last example is looping over a dataset to replace the null values with zeros.

Looping over a Sequence of Numbers

Listing 5-3 is to add the odd numbers only starting from 1 to 10 and print the sum and the number at every step.

Listing 5-3. Add the odd numbers from 0 to 10

```
#initialize sum with zero
sum=0
#for loop from 1-10
for num in range(1, 10):
    #check if it is an odd number
    if num%2!=0:
        #add the sum to the number
        sum=sum+num
```

```
#print both the number and the new summation
print('num=',num)
print('sum=',sum)
```

The for loop in Python has a different syntax than the other programming languages. The syntax is as follows:

```
for <element> in <iterable>:
```

Iterables in Python can be tuples, lists, strings, sequences, and more.

As in Listing 5-3, we start by initializing the variable sum with zero. Then, we use the for loop with the range() function to generate a sequence of numbers from 1 to 10. In the loop, we check if the number is odd. If it is true, we add the number to the sum and print both the number and the sum at every step for tracing. Pay attention that the range() function stops at the end number - 1. So, it starts from 1 and stops at 9, as in Figure 5-4. It does not take 10 into consideration.

```
num= 1
sum= 1
num= 3
sum= 4
num= 5
sum= 9
num= 7
sum= 16
num= 9
sum= 25
```

Figure 5-4. *Adding the odd numbers from 1 to 10*

Looping over a String

In the next two examples, we loop over the letters in a string and print the string's letters in reverse order. There are many ways to do this task; in this section, we see two of them. Like the previous example, we follow the for loop syntax using the in operator. In Listing 5-4, the string is "Python."

We slice the string and use -1 in the step to reverse it. The slicing square brackets syntax is as follows: [start :end :step].

Listing 5-4. Print string's letters in a reverse order

```
for letter in 'Python'[::-1]:
    print(letter)
```

The output is Figure 5-5.

```
n
o
h
t
y
P
```

Figure 5-5. *Output of Listing 5-4*

Figure 5-5 shows that when we used the step=-1, it started from the last letter and reversed the word's letters. Also, it is worth mentioning that the print() function in Python prints each iteration on a new line by default. It is equivalent to println() function in C/C++ programming languages. Later in Listing 5-6, we will see how to print all the iterations' output on the same line.

In the following example, we re-use the range() function for printing the string in a reverse order. In Listing 5-5, we initialize a variable p with the string. Then, we loop in the range starting from the length of the string to zero with a step of -1 to reverse the order. Remember that the range loops from the start number to the end number -1; so, if you put 1 instead of zero, the loop won't print the first letter of the string. However, feel free to play with the values to understand how it works. The output is as Figure 5-5.

Listing 5-5. Print string's letters in a reverse order

```
p='Python'
for i in range(len(p),0,-1):
    print(p[i-1])
```

The output of Listing 5-5 is the same as the previous listing. Listing 5-4. Both will get the output in Figure 5-5.

Looping over a List

In this example, we create a list of words and print them. By default, the print() function in Python prints each value on a new line, as we mentioned before. To change form this style, let us print the words on the same line having a separator as "^" between the words, as in Listing 5-6.

Listing 5-6. For loop over a list

```
my_list=['I','Love',"Python"]
for c in my_list:
    print(c,end="^")
```

We start by initializing the list with words, as in Listing 5-6. In Python, you can use both single and double quotes for the strings. Then, we loop over the list of strings and print the words one by one. To make the printing on the same line, we use the end= option and set it to the delimiter "^", as shown in the output in Figure 5-6.

I^Love^Python^

Figure 5-6. *Output of looping over a list of strings*

Looping over a DataFrame

In the previous section in this chapter, we learned how to represent the NULL values in a dataset, so the Python interpreter reads it correctly. However, we have not discussed yet how to handle them. There are many ways in Python that help you to replace the missing values to avoid propagating them and to avoid producing wrong analytics reports.

In the following example, Listing 5-7, we use the `fillna()` method to fill in the missing values in the dataset. First, we start by importing the required libraries and loading the excel file. To verify that the dataset has been read correctly, we print it and add three empty lines using "\n\n\n" after the DataFrame to easily distinguish the before and after.

Then, we loop over all the columns of the DataFrame using its `columns` property. Inside the loop, we use the `fillna()` method of the DataFrame to replace the NaN values with zeros. It is entirely up to data scientists to decide how to replace the null values. You can replace it with the average or with maximum or the minimum values, and so on. The most important note is to document the new value and write in your analytics report your reasoning and justification for choosing such value. In the next chapter, we will discuss the procedure and the options in more detail.

Listing 5-7. For loop over a DataFrame

```
import pandas as pd
import numpy as np
import matplotlib.pyplot as plt

students_df = pd.read_excel('../Datasets/Chapter 5/missing_
null.xlsx')

print(students_df,'\n\n\n')
```

```
for i in students_df.columns:
    students_df[i]=students_df[i].fillna(0)
print(students_df)
```

The output of the listing is in Figure 5-7.

```
    No      Name     Height      Weight    Age
0    0   Student1  164.981605  72.533380  16.0
1    1   Student2        NaN  51.277224   NaN
2    2   Student3  179.279758  58.146509  22.0
3    3   Student4  188.796394        NaN  24.0
4    4   Student5        NaN  65.523149  20.0
5    5   Student6  156.239781  80.332918  23.0
6    6   Student7        NaN  49.395245   NaN
7    7   Student8  184.647046  68.140550  21.0
8    8   Student9  174.044600        NaN  19.0
9    9  Student10  152.323344  53.985320  10.0
10  10  Student11  170.990257  72.339518  19.0
11  11  Student12        NaN  52.673586  23.0
12  12  Student13  183.297706  47.927322  23.0
13  13  Student14        NaN  87.699849  23.0
14  14  Student15  161.649166  64.806862  17.0
15  15  Student16        NaN  81.377881   NaN
16  16  Student17  162.169690        NaN  23.0
17  17  Student18        NaN  47.090269   5.0
18  18  Student19  167.277801  75.790486  24.0
19  19  Student20        NaN  61.486283  11.0


    No      Name     Height      Weight    Age
0    0   Student1  164.981605  72.533380  16.0
1    1   Student2    0.000000  51.277224   0.0
2    2   Student3  179.279758  58.146509  22.0
3    3   Student4  188.796394    0.000000  24.0
4    4   Student5    0.000000  65.523149  20.0
5    5   Student6  156.239781  80.332918  23.0
6    6   Student7    0.000000  49.395245   0.0
7    7   Student8  184.647046  68.140550  21.0
8    8   Student9  174.044600    0.000000  19.0
9    9  Student10  152.323344  53.985320  10.0
10  10  Student11  170.990257  72.339518  19.0
11  11  Student12    0.000000  52.673586  23.0
12  12  Student13  183.297706  47.927322  23.0
13  13  Student14    0.000000  87.699849  23.0
14  14  Student15  161.649166  64.806862  17.0
15  15  Student16    0.000000  81.377881   0.0
16  16  Student17  162.169690    0.000000  23.0
17  17  Student18    0.000000  47.090269   5.0
18  18  Student19  167.277801  75.790486  24.0
19  19  Student20    0.000000  61.486283  11.0
```

Figure 5-7. *The dataset before and after handling the missing values*

Summary

This chapter starts with how to make a comment in Python programs and how to use the arithmetic, comparison, logical, membership, and identity operators. Further, we discussed the different values that you can use to represent missing values in your datasets, so the Python interpreter reads them correctly. Moreover, we discussed the loops in detail with plenty of examples. There are the while and for loops. At the end of the chapter, we learned one of the ways of how to handle the missing values in Python. In the next chapter, we shall learn how to prepare for analysis and learn the essential conditional statement.

## CHAPTER 6

Preparing Data for Analysis

As we mentioned earlier, in any programming language, you learn the data types, operators, IF condition, and loops. We covered all these topics in Chapters 2 and 5, except the IF condition statements. In this chapter, we learn about them and more advanced data preparation and processing techniques.

Rename

In data analysis, an essential first step is to understand the variables of your dataset. You need to identify which variables are dependent and independent, as well as distinguish between numeric/continuous and character variables. Typically, your client provides this information in a text file accompanying the dataset. This text file, known as the dictionary, explains the variable names and their types. For instance, without the dictionary, variable names like YOB, ENROLL, and DT_ACCEPT might be unclear. The dictionary clarifies that YOB stands for Year Of Birth.

To add these descriptive labels to a dataset, we use a process of renaming columns. The dataset is in the Datasets Folder and is named Voter_A.CSV. After uploading and importing the file, we rename the dataset to Voters. Note that this dataset is synthetic, created using the variable names provided by the Secretary of Maine, but does not contain

© Engy Fouda 2024

E. Fouda, *Learn Data Science Using Python*, https://doi.org/10.1007/979-8-8688-0935-4_6

any real voters' information. I generated random 100 rows of voters' information for these exercises.

After importing, we see that the dataset contains 100 rows and 9 columns. We then rename the columns to include descriptive labels, making the dataset more readable. However, the data clearly needs cleaning. For instance, the Special Designations column is blank, and other columns have missing values. Verifying the data's validity and ensuring it is error-free is essential. We will learn how to clean such messy data in the remaining sections.

Listing 6-1. Rename method

```
import pandas as pd

df = pd.read_csv('../Datasets/Chapter 6/Voters_A.csv')
print(df.head())
# Renaming the columns to add labels
df = df.rename(columns={
    'FIRST_NAME': 'Name',
    'YOB': 'Year Of Birth',
    'ENROLL': 'Enrollment Code',
    'DESIGNATE': 'Special Designations',
    'DT_ACCEPT': 'Date Accepted (Date of Registration)',
    'CG': 'Congressional District',
    'CTY': 'County ID',
    'DT_CHG': 'Date Changed',
    'DT_LAST_VPH': 'Date Of Last Statewide Election with VPH'
})

# Displaying the DataFrame
print(df)
```

```
# Alternatively, if you need to display column names clearly
print("DataFrame with labeled columns:")
for column in df.columns:
    print(column)

# Save the modified DataFrame if needed
# df.to_csv('voters_labeled.csv', index=False)
```

In Listing 6-1, the `pd.read_csv('Voter_A.csv')` function loads the dataset into a Pandas DataFrame. The `.rename()` method assigns descriptive labels to the columns. To display the DataFrame with the new labels, we use `print(df)` method. The first step in cleaning the data is handling missing values by filling them with "Unknown" or "Missing." The output is in Figure 6-1.

Optionally, we can save the cleaned DataFrame to a new CSV file. This Python code achieves the goals of understanding and renaming variables for better readability and performing basic data cleaning.

```
            FIRST NAME   YOB ENROLL DESIGNATE    DT ACCEPT CG    CTY      DT CHG  \
0        Eula Vitolo   1913       R             12/5/1935  2  01AND  11/26/2008
1      Walton Santoyo  1918       R              9/8/1947  2  01AND   5/22/2008
2      Aletha Stabile  1925       D            10/14/1952  2  01AND   5/17/2010
3      Nannette Thong  1928       D             11/8/2005  2  01AND   4/25/2012
4     Courtney Bonner  1929       R            10/20/2009  2  01AND   6/13/2012

   DT LAST VPH
0
1   6/10/2008
2   11/2/2010
3   11/4/2008
4   6/14/2016
            FIRST NAME Year Of Birth Enrollment Code Special Designations  \
0        Eula Vitolo           1913                R
1      Walton Santoyo          1918                R
2      Aletha Stabile          1925                D
3      Nannette Thong          1928                D
4     Courtney Bonner          1929                R
..              ...            ...                ...                  ...
95        Erick Roth          1981                U
96    Francesco Pharis        1981                U
97         Eddy Shoop         1981                U
98     Quintin Speegle        1982                D
99     Mervin Margolis        1982                D

      DT ACCEPT Congressional District County ID      DT CHG DT LAST VPH
0     12/5/1935                      2    01AND  11/26/2008
1      9/8/1947                      2    01AND   5/22/2008   6/10/2008
2    10/14/1952                      2    01AND   5/17/2010  11/2/2010
3     11/8/2005                      2    01AND   4/25/2012  11/4/2008
4    10/20/2009                      2    01AND   6/13/2012  6/14/2016
..          ...                    ...      ...         ...         ...
95    4/28/2004                      2    01AND   3/11/2011
96    10/4/2004                      2    01AND  11/14/2006  11/4/2014
97    11/4/2008                      2    01AND  12/30/2009  11/4/2008
98     3/7/2000                      2    01AND  12/31/2005  11/6/2012
99    11/7/2006                      2    01AND  11/12/2008  11/6/2012

[100 rows x 9 columns]
DataFrame with labeled columns:
FIRST NAME
Year Of Birth
Enrollment Code
Special Designations
DT ACCEPT
Congressional District
County ID
DT CHG
DT LAST VPH
```

Figure 6-1. *Output with the new variable names*

Format

One of the most popular data exercises in healthcare is calculating the time a patient passes at a clinic or hospital and, based on historical data, making predictions for the patients' arrival times to decrease the waiting time. In this example, Listing 6-2, we make up a dataset with patients' ID, age, arrival date, arrival time, leaving date, leaving time, and visit fees. Then, we compute the time the patients passed inside the clinic and print a report of this duration and the money with dollar sign and commas in the thousands. This example aims to learn how Python handles date and time processing and currency formatting.

Listing 6-2. Date and currency formatting

```
import pandas as pd
import pandas as pd
from datetime import datetime

# Sample DataFrame
data = {
    'PatientID': [1, 2, 3],
    'Age': [25, 34, 50],
    'ArrivingDate': ['2024-06-10', '2024-06-11', '2024-06-12'],
    'ArrivingTime': ['09:00', '10:30', '14:45'],
    'LeavingDate': ['2024-06-10', '2024-06-11', '2024-06-12'],
    'LeavingTime': ['10:00', '11:15', '15:30'],
    'VisitFees': [100.0, 150.75, 200.5]
}

df = pd.DataFrame(data)
print("Datatypes of raw data:")
print("\n",df.dtypes)
```

```python
print("\nHeader of the original dataset:(",len(df.
columns),"columns)")
print("\n",df.head())

# Convert date and time columns to datetime
df['ArrivingDateTime'] = pd.to_datetime(df['ArrivingDate'] + ' '
+ df['ArrivingTime'])
df['LeavingDateTime'] = pd.to_datetime(df['LeavingDate'] + ' '
+ df['LeavingTime'])

# Calculate the duration of the visit
df['VisitDuration'] = df['LeavingDateTime'] - df
['ArrivingDateTime']

dummy=df[['VisitDuration', 'LeavingDateTime',
'ArrivingDateTime']]
print("\nComputing the visit duration:")
print("\n",dummy.head())

# Function to format the duration
def format_duration(duration):
    seconds = duration.total_seconds()
    hours = int(seconds // 3600)
    minutes = int((seconds % 3600) // 60)
    return f"{hours}h {minutes}m"

# Function to format the fees
def format_fees(fees):
    return '${:,.2f}'.format(fees)

# Print the report
print("\nClinic Visit Report:")

# display the DataFrame with formatted columns for reference
```

```
df['FormattedDuration'] = df['VisitDuration'].apply(format_
duration)
df['FormattedFees'] = df['VisitFees'].apply(format_fees)
print("\n", df[['PatientID', 'Age', 'FormattedDuration',
'FormattedFees']])
```

Let us explain the code in Listing 6-2. First, we create a DataFrame with columns: PatientID, Age, ArrivingDate, ArrivingTime, LeavingDate, LeavingTime, and VisitFees, and initialize it with dummy values.

Then, we use the pd.to_datetime() method to combine ArrivingDate and ArrivingTime into a single ArrivingDateTime column. Similarly, combine LeavingDate and LeavingTime into a LeavingDateTime column. To calculate the duration of the visit, subtract ArrivingDateTime from LeavingDateTime to compute the VisitDuration. We will next write a function to format this VisitDuration variable.

To format the duration, we define a function using the def keyword that takes the duration as an argument and returns it formatted. The format_duration(duration) converts the duration into a string format of hours and minutes, as in the output in Figure 6-2.Please check the following link to learn more about formatting dates: https://www.w3schools.com/python/python_datetime.asp.

About formatting the money currency, we define another function, format_fees(fees), to format the fees with a dollar sign and commas using '${:,.2f}'.format(fees), where $ is to set the currency symbol, :, is for the thousand's separator, and .2f is for the floating-point precision, setting it to two decimal places. Here is a link to learn more about currency formatting: https://www.geeksforgeeks.org/how-to-format-numbers-as-currency-strings-in-python/.

Finally, we specify the columns to print the report, as in Figure 6-2.

```
Datatypes of raw data:

 PatientID          int64
Age                 int64
ArrivingDate        object
ArrivingTime        object
LeavingDate         object
LeavingTime         object
VisitFees           float64
dtype: object

Header of the original dataset:( 7 columns)

      PatientID  Age ArrivingDate ArrivingTime LeavingDate LeavingTime   VisitFees
0             1   25   2024-06-10        09:00  2024-06-10       10:00      100.00
1             2   34   2024-06-11        10:30  2024-06-11       11:15      150.75
2             3   50   2024-06-12        14:45  2024-06-12       15:30      200.50

Computing the visit duration:

       VisitDuration      LeavingDateTime      ArrivingDateTime
0 0 days 01:00:00 2024-06-10 10:00:00 2024-06-10 09:00:00
1 0 days 00:45:00 2024-06-11 11:15:00 2024-06-11 10:30:00
2 0 days 00:45:00 2024-06-12 15:30:00 2024-06-12 14:45:00

Clinic Visit Report:

    PatientID  Age FormattedDuration FormattedFees
0           1   25            1h 0m        $100.00
1           2   34            0h 45m       $150.75
2           3   50            0h 45m       $200.50
```

Figure 6-2. *Report of durations patients passed at the venue*

Creating New Variables

In Python, you do not need to predefine your variables. In this example, we will continue working with the presidential elections project and the Voters_A.csv dataset. The dataset has a variable called YOB, which is the Year Of Birth. From the voters' groupings is the age. In Listing 6-3, we will compute the voters' ages by subtracting the YOB from the current year.

Listing 6-3. Creating new variables

```python
import pandas as pd

df = pd.read_csv('../Datasets/Chapter 6/Voters_A.csv')
print("Dataframe Shape (rows, columns):",df.shape)
print("\nRaw data-first 5 rows:\n",df.head())
print("\nDatatypes of raw data:\n")
print("\n",df.dtypes)

#Typecasting-converting to numeric
df['YOB_numeric'] = pd.to_numeric(df['YOB'], errors='coerce')
print("\nDatatypes after typecasting:\n")
print("\n",df.dtypes)
# Calculate age using 'YOB_numeric'
age = 2024 - df['YOB_numeric']

print("\nage:\n",age.head())

df['age']=age
print("\nDataframe's first 5 rows after creating new
variables:\n",df.head())
print("\nDataframe Shape after adding new columns to it (rows,
columns):",df.shape)
```

As always, after loading the libraries and the dataset, you verify that it
has been loaded correctly by printing the dataset dimensions using df.
shape(), the first five rows using the df.head() method, and the datatypes
of the variables using the df.types property to check if we need any
typecasting before performing the mathematical operations.

As shown in Figure 6-3, the dimensions of the dataset are 100 rows
and 9 columns. However, the datatypes of all the columns are object. This
output indicates that there are empty strings or non-numeric values in
the YOB column that cannot be directly converted to integers. Therefore,
we will need to change the datatype of the YOB to numeric to be able to

perform an arithmetic operation over it. As in Listing 6-3, we use the pd. to_numeric() method to do the typecasting with the errors='coerce' parameter, which will convert invalid parsing to NaN. This identification allows us to fill these NaN values with a default value before calculating the age. However, in this exercise, we will leave it as NaN.

If you would like to fill the NaN values in YOB with a default value (e.g., 2006 to get the minimum voting age of 18 years old or any appropriate value), you can use df['YOB'].fillna(2006, inplace=True). It will replace the NaN in place.

Afterward, we print the datatypes again to check the new variable YOB_numeric's datatype. As in Figure 6-3, it is float64. Now, it is a numeric datatype, and we can perform the subtraction to compute age. Then, we print the first five rows of age using age.head() method. To add this Series as a new column to the DataFrame df, we use df['age']=age. For final verification, we printed the head and the dimensions of the DataFrame to find that the columns increased from 9 to 11.

```
Dataframe Shape (rows, columns): (100, 9)

Raw data-first 5 rows:
            FIRST NAME   YOB ENROLL DESIGNATE   DT ACCEPT CG     CTY       DT CHG  \
0        Eula Vitolo   1913      R             12/5/1935  2   01AND   11/26/2008
1      Walton Santoyo   1918      R              9/8/1947  2   01AND    5/22/2008
2      Aletha Stabile   1925      D            10/14/1952  2   01AND    5/17/2010
3      Nannette Thong   1928      D             11/8/2005  2   01AND    4/25/2012
4     Courtney Bonner   1929      R            10/20/2009  2   01AND    6/13/2012

    DT LAST VPH
0
1    6/10/2008
2   11/2/2010
3   11/4/2008
4   6/14/2016

Datatypes of raw data:


 FIRST NAME      object
YOB             object
ENROLL          object
DESIGNATE       object
DT ACCEPT       object
CG              object
CTY             object
DT CHG          object
DT LAST VPH     object
dtype: object

Datatypes after typecasting:


 FIRST NAME      object
YOB             object
ENROLL          object
DESIGNATE       object
DT ACCEPT       object
CG              object
CTY             object
DT CHG          object
DT LAST VPH     object
YOB_numeric    float64
dtype: object

age:
 0    111.0
1    106.0
2     99.0
3     96.0
4     95.0
Name: YOB_numeric, dtype: float64

Dataframe's first 5 rows after creating new variables:
            FIRST NAME   YOB ENROLL DESIGNATE   DT ACCEPT CG     CTY       DT CHG  \
0        Eula Vitolo   1913      R             12/5/1935  2   01AND   11/26/2008
1      Walton Santoyo   1918      R              9/8/1947  2   01AND    5/22/2008
2      Aletha Stabile   1925      D            10/14/1952  2   01AND    5/17/2010
3      Nannette Thong   1928      D             11/8/2005  2   01AND    4/25/2012
4     Courtney Bonner   1929      R            10/20/2009  2   01AND    6/13/2012

    DT LAST VPH  YOB_numeric    age
0                      1913.0  111.0
1    6/10/2008         1918.0  106.0
2   11/2/2010          1925.0   99.0
3   11/4/2008          1928.0   96.0
4   6/14/2016          1929.0   95.0

Dataframe Shape after adding new columns to it (rows, columns): (100, 11)
```

Figure 6-3. *Compute the voters' ages*

Rearrange the Dataset Variables

To rearrange the variables to get the age variable directly beside the YOB column for easy checking, we use the name of the source DataFrame followed by two square brackets and then list the columns that you want to keep in the order that you specify separated by commas and each variable name between quotations.

In Listing 6-4, after we import the dataset and compute the age column, as we explained in the previous section, we rearrange the DataFrame in place by listing the age after the YOB. We print the first five rows to verify the change using the df.head() method.

Listing 6-4. Rearrange the dataset variables

```
import pandas as pd
df = pd.read_csv('../Datasets/Chapter 6/Voters_A.csv')
print("\nRaw data-first 5 rows:\n",df.head())
df['YOB_numeric'] = pd.to_numeric(df['YOB'], errors='coerce')

# Calculate age using 'YOB_numeric'
df['age'] = 2024 - df['YOB_numeric']

df=df[['FIRST NAME', 'YOB', 'age', 'ENROLL', 'DESIGNATE','DT
ACCEPT', 'CG', 'CTY','DT CHG', 'DT LAST VPH']]
print("\nThe df after rearranging:\n",df.head())
```

The output is as in Figure 6-4.

```
Raw data-first 5 rows:
          FIRST NAME   YOB ENROLL DESIGNATE    DT ACCEPT CG    CTY    DT CHG  \
0       Eula Vitolo  1913      R               12/5/1935   2  01AND  11/26/2008
1     Walton Santoyo  1918     R                9/8/1947   2  01AND   5/22/2008
2     Aletha Stabile  1925     D              10/14/1952   2  01AND   5/17/2010
3     Nannette Thong  1928     D               11/8/2005   2  01AND   4/25/2012
4   Courtney Bonner   1929     R              10/20/2009   2  01AND   6/13/2012

   DT LAST VPH
0
1   6/10/2008
2  11/2/2010
3  11/4/2008
4   6/14/2016

The df after rearranging:
          FIRST NAME   YOB    age ENROLL DESIGNATE    DT ACCEPT CG    CTY  \
0       Eula Vitolo  1913  111.0      R               12/5/1935   2  01AND
1     Walton Santoyo  1918 106.0      R                9/8/1947   2  01AND
2     Aletha Stabile  1925  99.0      D              10/14/1952   2  01AND
3     Nannette Thong  1928  96.0      D               11/8/2005   2  01AND
4   Courtney Bonner   1929  95.0      R              10/20/2009   2  01AND

           DT CHG DT LAST VPH
0  11/26/2008
1   5/22/2008   6/10/2008
2   5/17/2010  11/2/2010
3   4/25/2012  11/4/2008
4   6/13/2012   6/14/2016
```

Figure 6-4. *Voters dataset after reordering its variables to have age after YOB*

IF Statement

The age column has a wrong value for one of the voters, where they are 344 years old. To correct this, if the dataset data was a real one, we would have had several options:

1. Obtain the correct age by contacting the Secretary of Maine's office or the voter directly.

2. Remove the value and put NaN instead of it to indicate that it is missing data, as mentioned in Chapter 5.

3. Delete the entire row containing the incorrect value.

4. Leave the value as is, though it would be an outlier
    and could skew the analysis of the relationship
    between age, party affiliation, and voting
    predictions.

In this example, we will make it a missing value, NaN. Later in this
chapter, we will learn how to delete the whole row.

In Listing 6-5, we start as in the previous section by loading the
libraries and the dataset, verifying the loading, and computing the age
column. Then, we loop over the rows using a for loop and use the df.
index property to return the maximum number of rows -1 because Python
counts from zero.

We use if df.loc[x, "age"] > 150 to check if the value in the
column age at index x is greater than 150. We print the row with the wrong
age value, then replace it with NaN, and reprint again using the index
location x using df.loc[x].

Listing 6-5. IF statement

```
import pandas as pd

df = pd.read_csv('../Datasets/Chapter 6/Voters_A.csv')
print("\nRaw data-first 5 rows:\n",df.head())
df['YOB_numeric'] = pd.to_numeric(df['YOB'], errors='coerce')

df['age'] = 2024 - df['YOB_numeric']

for x in df.index:
    if df.loc[x, "age"] > 150:
        print("\nThe rows with the wrong ages:\n",df.loc[x])
        df.loc[x, "age"] = "NaN"
        print("\nReplacing the wrong ages with NaN:
        \n",df.loc[x])
```

The output is in Figure 6-5. The location is printed as row 89 because Python counts from zero.

```
Raw data-first 5 rows:
           FIRST NAME   YOB ENROLL DESIGNATE   DT ACCEPT CG    CTY      DT CHG  \
0        Eula Vitolo  1913      R                12/5/1935  2  01AND  11/26/2008
1     Walton Santoyo  1918      R                 9/8/1947  2  01AND   5/22/2008
2      Aletha Stabile 1925      D               10/14/1952  2  01AND   5/17/2010
3     Nannette Thong  1928      D                11/8/2005  2  01AND   4/25/2012
4    Courtney Bonner  1929      R               10/20/2009  2  01AND   6/13/2012

    DT LAST VPH
0
1    6/10/2008
2   11/2/2010
3   11/4/2008
4    6/14/2016

The rows with the wrong ages:
 FIRST NAME      Mckinley Guynn
YOB                        1680
ENROLL                        D
DESIGNATE
DT ACCEPT          11/27/2007
CG                            2
CTY                      01AND
DT CHG             10/18/2011
DT LAST VPH        11/4/2014
YOB_numeric            1680.0
age                     344.0
Name: 89, dtype: object

Replacing the wrong ages with NaN:
 FIRST NAME      Mckinley Guynn
YOB                        1680
ENROLL                        D
DESIGNATE
DT ACCEPT          11/27/2007
CG                            2
CTY                      01AND
DT CHG             10/18/2011
DT LAST VPH        11/4/2014
YOB_numeric            1680.0
age                         NaN
Name: 89, dtype: object
```

Figure 6-5. *IF statement output*

IF-ELIF Statements

Now, we will analyze the number of voters in the following age groups: 18–24, 25–29, 30–39, 40–49, 50–64, and 65+.

Listing 6-6. IF-ELIF statements

```
import pandas as pd
import matplotlib.pyplot as plt

df = pd.read_csv('../Datasets/Chapter 6/Voters_A.csv')
print("\nRaw data-first 5 rows:\n",df.head())
df['YOB_numeric'] = pd.to_numeric(df['YOB'], errors='coerce')

df['age'] = 2024 - df['YOB_numeric']
for x in df.index:
    #if (df.loc[x,'age']>=18 and df.loc[x,'age']<=24):
    if (18<=df.loc[x,'age']<=24):
        df.loc[x,'age_group']='18-24'
    elif (25<=df.loc[x,'age']<=29) :
        df.loc[x,'age_group']='25-29'
    elif (30<=df.loc[x,'age']<=39) :
        df.loc[x,'age_group']='30-39'
    elif (40<=df.loc[x,'age']<=49) :
        df.loc[x,'age_group']='40-49'
    elif (50<=df.loc[x,'age']<=64) :
        df.loc[x,'age_group']='50-64'
    elif (df.loc[x,'age']>=65) :
        df.loc[x,'age_group']='65+';

print("\nchecking age_group:\n",df.head())

# Create frequency table directly from value_counts
frequency_table = df['age_group'].value_counts().to_
frame('Frequency').reset_index()

# Rename the columns to be consistent
frequency_table.columns = ['age_group', 'Frequency']
```

```
# Sort the table by frequency
frequency_table = frequency_table.sort_values(by='Frequency',
ascending=False)

# Print the frequency table
print("\n",frequency_table)

# Plot the frequency
plt.figure(figsize=(8, 6))
plt.bar(frequency_table['age_group'], frequency_
table['Frequency'], color='skyblue')
plt.xlabel('age_group')
plt.ylabel('Frequency')
plt.title('Frequency Plot of age_group')
plt.show()
```

In Listing 6-6, as always, we start by loading the libraries and the dataset and verifying that the dataset has been loaded correctly by printing the first five rows. Then, as we explained in previous sections of this chapter, we compute the age column.

To compute the age_group variable, we loop over the datasets row by row and made nested if-elif conditions using the age ranges, for example, 18<=df.loc[x,'age']<=24. Python allows using the ranges, and it also allows the compound conditional statements as other programming languages, df.loc[x,'age']>=18 and df.loc[x,'age']<=24. Also, you can use else if as in other programming languages instead of the elif statement.

Afterward, as we explained in Chapter 4, we compute the frequency table of the age_group and plot its distribution using a bar chart.

We can use many other ways to compute the age_group other than the nested if-elif statements. One of them is using the pd.cut() method. It would be more pythonic. However, this section aims to show how to use IF-ELIF statements. We can replace the whole section of the IF-ELIF statements with the following three lines:# Define age groups

```
age_bins = [18, 25, 30, 40, 50, 65, float('inf')]
# Define corresponding labels
age_labels = ['18-24', '25-29', '30-39', '40-49',
'50-64', '65+']
# Assign age groups using pd.cut
df['age_group'] = pd.cut(df['age'], bins=age_bins, labels=age_
labels, right=False)
```

To create a frequency table to the age_group and make a bar chart for it, we use to_frame() method to convert the Series returned by value_counts() directly into a DataFrame and then rename the columns consistently. Then, we rename the frequency_table columns' names by using the .columns property as in this statement: frequency_table. columns = ['age_group', 'Frequency'].

At the end of this code snippet, we print the frequency_table and plot it as a bar chart, using plt.bar(frequency_table['age_group'], frequency_table['Frequency'], color='skyblue'). Afterward, we set the labels of the x and y axes and the title. Finally, we show the plot with these settings. Figure 6-6 shows the output of Listing 6-6.

```
checking age_group:
          FIRST NAME   YOB ENROLL DESIGNATE    DT ACCEPT CG    CTY       DT CHG \
0       Eula Vitolo  1913     R               12/5/1935  2  01AND  11/26/2008
1    Walton Santoyo  1918     R                9/8/1947  2  01AND   5/22/2008
2    Aletha Stabile  1925     D              10/14/1952  2  01AND   5/17/2010
3    Nannette Thong  1928     D               11/8/2005  2  01AND   4/25/2012
4   Courtney Bonner  1929     R              10/20/2009  2  01AND   6/13/2012

    DT LAST VPH  YOB_numeric    age age_group
0                    1913.0  111.0       65+
1     6/10/2008      1918.0  106.0       65+
2    11/2/2010       1925.0   99.0       65+
3    11/4/2008       1928.0   96.0       65+
4     6/14/2016      1929.0   95.0       65+

    age_group  Frequency
0     40-49          49
1     50-64          33
2       65+          15
3     30-39           1
4     25-29           1
```

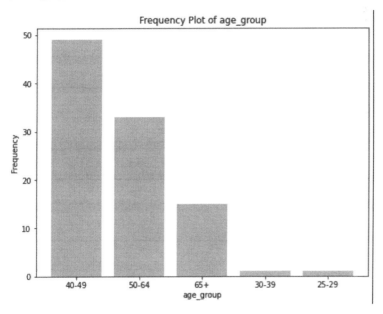

Figure 6-6. *Computing the age_group of the voters, its frequency table, and a histogram*

Drop a Row

As mentioned in the "IF Statement" section, there are various ways to handle the wrong data. In the previous section, we saw how to replace them with missing values. In this section, we will learn how to drop the whole raw.

Listing 6-7. Drop a row

```
import pandas as pd

df = pd.read_csv('../Datasets/Chapter 6/Voters_A.csv')
print("Dataframe Shape (rows, columns):",df.shape)
print("\nRaw data-first 5 rows:\n",df.head())

df['YOB_numeric'] = pd.to_numeric(df['YOB'], errors='coerce')

rows_with_errors = df[df['YOB_numeric'].isna()]
print("\nRows with coercing errors in 'YOB':")
print(rows_with_errors)

df.drop([83],axis=0,inplace=True)
rows_with_errors = df[df['YOB_numeric'].isna()]
print("\nChecking if the row with errors was deleted or not:")
print(rows_with_errors)
print("Further verification by checking the Dataframe Shape
(rows, columns):",df.shape)
```

Listing 6-7 starts by loading the libraries and the dataset, as always. Then, we print the dataset dimensions, shown in Figure 6-7, as 100 rows and 9 columns. Then, we print the first five rows using the df. head() method.

Afterward, we typecast YOB to a new numeric column using the pd. to_numeric() method. We subset the rows with errors using the .isna()

142

method. As shown in the output in Figure 6-7, Row 83 is almost blank. It only has the voter's name only, and the other columns are blank.

Therefore, we use the df.drop() method with the row number to drop the whole row. Then, we redo the verification again using the .isna() method, and now, the list is empty, as shown in Figure 6-7. Finally, we reprinted the DataFrame dimensions to find that the number of rows decreased by 1–99.

```
Dataframe Shape (rows, columns): (100, 9)

Raw data-first 5 rows:
          FIRST NAME   YOB ENROLL DESIGNATE   DT ACCEPT CG    CTY      DT CHG  \
0       Eula Vitolo  1913      R               12/5/1935  2  01AND  11/26/2008   |
1     Walton Santoyo  1918      R                9/8/1947  2  01AND   5/22/2008
2     Aletha Stabile  1925      D              10/14/1952  2  01AND   5/17/2010
3     Nannette Thong  1928      D               11/8/2005  2  01AND   4/25/2012
4    Courtney Bonner  1929      R              10/20/2009  2  01AND   6/13/2012

   DT LAST VPH
0
1    6/10/2008
2   11/2/2010
3   11/4/2008
4    6/14/2016

Rows with coercing errors in 'YOB':
        FIRST NAME YOB ENROLL DESIGNATE DT ACCEPT CG CTY DT CHG DT LAST VPH  \
83  Cletus Bodine

       YOB_numeric
83           NaN

Checking if the row with errors was deleted or not:
Empty DataFrame
Columns: [FIRST NAME, YOB, ENROLL, DESIGNATE, DT ACCEPT, CG, CTY, DT CHG, DT LAST VPH, YOB_numeric]
Index: []
Further verification by checking the Dataframe Shape (rows, columns): (99, 10)
```

Figure 6-7. *Drop a row*

Drop a Column

The DROP method specifies excluding some variables from the dataset. In the Voters_A.csv dataset, the Designate column is empty. In cleaning the data, we should delete this column; in other words, drop it, as in Listing 6-8.

Listing 6-8. Drop method

```
import pandas as pd

import numpy as np
df = pd.read_csv('../Datasets/Chapter 6/Voters_A.csv')
print("The size of the dataframe:",df.shape)
df.replace(to_replace=[r'^\s*$', '', None], value=np.nan,
regex=True, inplace=True)
print(df.isnull().sum())

print("The dataframe before dropping:")
print(df.head())

df.drop("DESIGNATE",axis = 1, inplace = True)
print("The size of the dataframe:",df.shape)
print("The dataframe after dropping:")
print(df.head())
```

After loading the file correctly using the pd.csv() method, we verified the DataFrame size, which is shown in Figure 6-8 as (100,9).

Sometimes, empty cells might not be read as empty strings (' ') but as None, whitespace characters, or other representations. To handle various types of empty values and replace them with NaN. Therefore, we can use df.replace(to_replace=[r'^\s*$', '', None], value=np.nan, regex=True, inplace=True). This method replaces all representations of empty values with NaN, including a RegEx (Regular Expression) pattern to match any string that consists only of whitespace characters, r'^\s*$'. Also, it will replace empty strings and None.

The df.isnull().sum() counts the NaN values in each column. The output, as in Figure 6-8, shows that all the rows in the Designate column are NaN. We verify that using the .head() method.

The df.drop("DESIGNATE", axis=1, inplace=True) removes the column named DESIGNATE from the DataFrame. The option axis=1 indicates that a column (not a row) is to be dropped. In Pandas, axis=0 refers to rows, and axis=1 refers to columns. The option inplace=True modifies the original DataFrame df directly and does not create a new DataFrame with the new values.

We print the DataFrame header and size before and after dropping the column to verify that the number of columns decreased from 9 to 8, as shown in Figure 6-8.

```
The size of the dataframe: (100, 9)
FIRST NAME        0
YOB               1
ENROLL            1
DESIGNATE       100
DT ACCEPT         1
CG                1
CTY               1
DT CHG            1
DT LAST VPH      34
dtype: int64
The dataframe before dropping:
        FIRST NAME   YOB ENROLL  DESIGNATE   DT ACCEPT CG    CTY     DT CHG  \
0       Eula Vitolo  1913     R        NaN    12/5/1935  2  01AND  11/26/2008
1    Walton Santoyo  1918     R        NaN     9/8/1947  2  01AND   5/22/2008
2     Aletha Stabile 1925     D        NaN   10/14/1952  2  01AND   5/17/2010
3     Nannette Thong 1928     D        NaN    11/8/2005  2  01AND   4/25/2012
4    Courtney Bonner 1929     R        NaN   10/20/2009  2  01AND   6/13/2012

    DT LAST VPH
0         NaN
1   6/10/2008
2   11/2/2010
3   11/4/2008
4   6/14/2016
The size of the dataframe: (100, 8)
The dataframe after dropping:
        FIRST NAME   YOB ENROLL    DT ACCEPT CG    CTY     DT CHG DT LAST VPH
0       Eula Vitolo  1913     R    12/5/1935  2  01AND  11/26/2008        NaN
1    Walton Santoyo  1918     R     9/8/1947  2  01AND   5/22/2008  6/10/2008
2     Aletha Stabile 1925     D   10/14/1952  2  01AND   5/17/2010  11/2/2010
3     Nannette Thong 1928     D    11/8/2005  2  01AND   4/25/2012  11/4/2008
4    Courtney Bonner 1929     R   10/20/2009  2  01AND   6/13/2012  6/14/2016
```

Figure 6-8. *Eight columns after dropping one*

Subset

There are multiple ways to subset the Pandas DataFrames in Python. One of them is Boolean indexing. In this example, we continue using the dataset of the previous sections, Voters_A.CSV. We want to create two new datasets as subsets of the original, where the first one contains only voters who are enrolled as Democrats and the other includes voters who are enrolled as Republicans.

Listing 6-9. Subset the dataset

```
import pandas as pd

import numpy as np
df = pd.read_csv('../Datasets/Chapter 6/Voters_A.csv')

democrats=df[df['ENROLL']=="D"]
print(democrats.iloc[:,:6].head())
print(democrats.shape)

republicans=df[df['ENROLL']=="R"]
print(republicans.iloc[:,:6].head())
print(republicans.shape)

frequency_table = df['ENROLL'].value_counts().reset_index().
rename(columns={'index': 'ENROLL', 'ENROLL': 'Frequency'})
frequency_table = frequency_table.sort_values(by='Frequency',
ascending=False)
print(frequency_table)
```

In Listing 6-9, after importing the library and loading the dataset, we use df[df['ENROLL'] == "D"] to filter the DataFrame to include only the rows where the ENROLL column is equal to "D." This subset is stored in the democrats DataFrame.

To print the first six columns of the `democrats` DataFrame, we use `democrats.iloc[:, :6].head()`. The `iloc` property is used for indexing rows and columns. We used it here to select the first six columns. To get the number of Democratic voters in this dataset, we use `democrats.shape`.

We repeat the same steps for the Republican party. Then, at the end, we print the frequency table to compute all the unique levels of the `ENROLL` column and the number of the frequency of each level in this dataset. We use `df['ENROLL'].value_counts()` to calculate the frequency of each unique value in the `ENROLL` column. The `reset_index()` method converts the resulting Series into a DataFrame and resets the index. The `rename(columns={'index': 'ENROLL', 'ENROLL': 'Frequency'})` method is to rename the columns to ENROLL and Frequency for clarity. Finally, the method `sort_values(by='Frequency', ascending=False)` sorts the DataFrame by the `Frequency` column in descending order.

Figure 6-9 shows the first five rows of both smaller datasets, and the frequency table shows how many times each enrollment type appears in the dataset.

	FIRST NAME	YOB	ENROLL	DESIGNATE	DT ACCEPT	CG
2	Aletha Stabile	1925	D		10/14/1952	2
3	Nannette Thong	1928	D		11/8/2005	2
8	Trisha Tuch	1949	D		5/8/2006	2
9	Vonda Delahoussaye	1950	D		11/2/1988	2
11	Flavia Adolphson	1986	D		6/16/2012	2

(24, 9)

	FIRST NAME	YOB	ENROLL	DESIGNATE	DT ACCEPT	CG
0	Eula Vitolo	1913	R		12/5/1935	2
1	Walton Santoyo	1918	R		9/8/1947	2
4	Courtney Bonner	1929	R		10/20/2009	2
5	Leda Glessner	1932	R		10/21/1968	2
6	Cammie Sidoti	1935	R		12/16/2004	2

(24, 9)

	ENROLL	Frequency
0	U	47
1	R	24
2	D	24
3	G	3
4	L	1
5		1

Figure 6-9. Subset output and the frequency table

Append

Appending datasets is the same as vertical merging, which means adding rows. In Python, you can merge DataFrames both vertically and horizontally using the pandas library. In this section, we will append the two datasets, democrats and republicans, datasets that we created in the previous section. In the next section, we will have an example of horizontal merging.

You must run Listing 6-9 first to load the Democratic and Republican datasets before appending them using Listing 6-10.

Listing 6-10. Append two datasets

```
import pandas as pd

# Concatenate DataFrames vertically
vertical_merged_df = pd.concat([democrats,republicans
], axis=0)

print("The size of the new dataframe=",vertical_merged_
df.shape)
print("Vertical Merge Result:")
print(vertical_merged_df.iloc[:,:6])
```

In Listing 6-10, we use the pd.concat() method to merge DataFrames vertically; in other words, stacking them on top of each other. The option of axis=0 specifies that the concatenation should be performed along rows. This means rows from the republicans DataFrame are added to the rows of the democrats DataFrame, resulting in a new DataFrame that contains all rows from both DataFrames. The new latest dataset that has rows of both datasets is called vertical_merged_df.

To verify that the rows of the two datasets are in the new one, we check its dimensions by using the vertical_merged_df.shape property. It returns a tuple representing the dimensions of the DataFrame. It shows the dataset size, number of rows, and columns. Figure 6-10 shows that the size is (48,9) indicating that the append occurred correctly because each has 24 rows, as was shown in the frequency table in the previous section.

To make the output look nice, we display the first six columns only to have one line header and avoid confusion by some columns continuing on the following lines. Therefore, we use vertical_merged_df.iloc[:, :6] to select the first six columns of the concatenated DataFrame using the iloc method. It uses integer-based indexing, where : indicates all rows and :6 indicates the first six columns. So, we print the selected subset of the DataFrame, displaying the first six columns only for all rows in the new DataFrame.

149

```
The size of the new dataframe= (48, 9)
Vertical Merge Result:
```

	FIRST NAME	YOB	ENROLL DESIGNATE	DT ACCEPT	CG
2	Aletha Stabile	1925	D	10/14/1952	2
3	Nannette Thong	1928	D	11/8/2005	2
8	Trisha Tuch	1949	D	5/8/2006	2
9	Vonda Delahoussaye	1950	D	11/2/1988	2
11	Flavia Adolphson	1986	D	6/16/2012	2
12	Jarrod Sells	1995	D	12/8/2015	2
18	Carina Benson	1955	D	8/10/1998	2
29	Kylie Wolfgram	1971	D	12/4/1991	2
34	Salena Mcgowin	1972	D	01/01/1850	2
43	Alejandra Schuyler	1974	D	01/01/1850	2
44	Misti Shibata	1974	D	10/13/1992	2
45	Arleen Ranson	1974	D	11/5/1997	2
58	Walker Shippy	1976	D	8/30/1994	2
59	Jermaine Elston	1976	D	10/28/2004	2
60	Ellis Mcdaniel	1976	D	11/2/2004	2
73	Seymour Seiber	1978	D	9/25/2002	2
74	Waldo Weidman	1978	D	10/12/2004	2
75	Marco Mccarley	1978	D	11/2/2004	2
85	Zane Kime	1979	D	10/25/2008	2
86	Barney Xie	1979	D	11/4/2008	2
89	Mckinley Guynn	1680	D	11/27/2007	2
93	Benjamin Wilton	1981	D	2/23/2016	2
98	Quintin Speegle	1982	D	3/7/2000	2
99	Mervin Margolis	1982	D	11/7/2006	2
0	Eula Vitolo	1913	R	12/5/1935	2
1	Walton Santoyo	1918	R	9/8/1947	2
4	Courtney Bonner	1929	R	10/20/2009	2
5	Leda Glessner	1932	R	10/21/1968	2
6	Cammie Sidoti	1935	R	12/16/2004	2
7	Regina Bundy	1944	R	8/5/1971	2
16	Verena Sapienza	1938	R	9/28/1964	2
21	Kurtis Howle	1967	R	11/1/2004	2
25	Kylee Admire	1970	R	6/12/2006	2
30	Nguyet Scheider	1971	R	9/24/1992	2
38	Mi Henegar	1973	R	11/5/2002	2
47	Chanel Capra	1974	R	11/3/1998	2
51	Edmund Ballantine	1975	R	6/9/1998	2
67	Jackson Fallen	1977	R	2/10/2008	2
68	Numbers Wnuk	1977	R	11/5/1996	2
69	Tommie Brandi	1977	R	11/13/1995	2
77	Fidel Keown	1978	R	3/19/2004	2
78	Art Destefano	1978	R	10/25/2004	2
79	Mack Esses	1978	R	11/2/2004	2
80	Elmer Prestwood	1978	R	11/7/2006	2
87	Miguel Reading	1979	R	10/23/2012	2
90	Gregory Scherer	1980	R	10/18/2004	2
91	Rob Heth	1980	R	11/4/2014	2
94	Stephen Shaughnessy	1981	R	2/18/2015	2

Figure 6-10. *Appending two datasets in one*

You can also use `pd.concat()` to merge DataFrames horizontally (side by side) by setting the `axis=1`.

Merge

In the previous section, we learned how to append two datasets and increase the rows. In this section, we will learn how to merge two datasets over one common variable, where one has many variables and the other has a few ones.

In this example, we will use the datasets of a Kaggle competition posted by Zillow to make house predictions. The contest has two datasets. The Many dataset is loaded from a CSV file called `properties_2016.csv`, which contains 58 columns and about three million rows. The Few dataset is loaded from another CSV file called `train_2016_v2.csv`, which contains three columns and more than 90K rows. The common variable between both files is `parcelid`. You can download the original datasets from this link: `https://www.kaggle.com/competitions/zillow-prize-1/data`.

As the sizes of these files are enormous, I selected only 3000 rows from the first file and 1000 rows from the second file and named them `many.csv` and `few.csv`. You will find them in the Datasets Folder.

Listing 6-11. Merge a few-to-many datasets

```
import pandas as pd

file1 = pd.read_csv('../Datasets/Chapter 6/many.csv',
low_memory=False)
file2 = pd.read_csv('../Datasets/Chapter 6/few.csv',
low_memory=False)
print("\nThe Many Dataset Dimensions:\n",file1.shape)
print("\nThe Few Dataset Dimensions:\n",file2.shape)
```

```
zillow_merged = pd.merge(file1, file2, on='parcelid')
print("\nThe Merged Dataset dimensions:\n",zillow_merged.shape)

print("\nThe Merged Dataset Head:\n",zillow_merged.head())
```

In Listing 6-11, we use low_memory=False option when loading the datasets to avoid chunking. Setting this parameter to False makes Pandas read the entire file into memory before parsing it. This can be used when columns have mixed types by allowing Pandas to infer the correct data types more accurately. You can also manually inspect and convert the columns after loading the data by looping and fixing the mixed datatypes of the columns that cause errors. However, this is beyond the scope of this exercise, so that we will use the parameter, and that is it.

To concatenate the DataFrames horizontally on a key variable, we use the pd.merge() method and specify parcelid as the key variable in the on parameter. By default, the merge method uses Inner Join. So, it only keeps the rows where the parcelid exists in file1 and file2. This is why the resulting DataFrame has the number of rows close to file2, which is 1004. The extra four rows are due to duplicates in the parcelid.

You can remove the duplicates by using these statements:

```
file1 = file1.drop_duplicates(subset='parcelid')
file2 = file2.drop_duplicates(subset='parcelid')
```

If you want to include all rows from file1 regardless of whether there is a matching parcelid in file2, you can use several types of joins, such as a Left Join. The statement will be zillow_merged = pd.merge(file1, file2, on='parcelid', how='left'). You will have NaN for the nonmatching columns. The merged DataFrame, in such case, will have about the number of rows as the Many file, which is 3000 rows.

Merging in Python is easy and does not require sorting by the key identifier as in other languages.

The output is shown in Figure 6-11.

```
The Many Dataset Dimensions:
 (3000, 58)

The Few Dataset Dimensions:
 (1000, 3)

The Merged Dataset dimensions:
 (1004, 60)

The Merged Dataset Head:
    parcelid  airconditioningtypeid  architecturalstyletypeid  basementsqft  \
0  17073783                    NaN                       NaN           NaN
1  17088994                    NaN                       NaN           NaN
2  17100444                    NaN                       NaN           NaN
3  17102429                    NaN                       NaN           NaN
4  17109604                    NaN                       NaN           NaN

   bathroomcnt  bedroomcnt  buildingclasstypeid  buildingqualitytypeid  \
0         2.5         3.0                  NaN                    NaN
1         1.0         2.0                  NaN                    NaN
2         2.0         3.0                  NaN                    NaN
3         1.5         2.0                  NaN                    NaN
4         2.5         4.0                  NaN                    NaN

   calculatedbathnbr  decktypeid  ...  structuretaxvaluedollarcnt  \
0                2.5         NaN  ...                     115087.0
1                1.0         NaN  ...                     143809.0
2                2.0         NaN  ...                      33619.0
3                1.5         NaN  ...                      45609.0
4                2.5         NaN  ...                     277000.0

   taxvaluedollarcnt  assessmentyear  landtaxvaluedollarcnt  taxamount  \
0           191811.0          2015.0                76724.0    2015.06
1           239679.0          2015.0                95870.0    2581.30
2            47853.0          2015.0                14234.0     591.64
3            62914.0          2015.0                17305.0     682.78
4           554000.0          2015.0               277000.0    5886.92

   taxdelinquencyflag  taxdelinquencyyear  censustractandblock  logerror  \
0                 NaN                 NaN         6.111002e+13    0.0953
1                 NaN                 NaN         6.111002e+13    0.0198
2                 NaN                 NaN         6.111001e+13    0.0060
3                 NaN                 NaN         6.111001e+13   -0.0566
4                 NaN                 NaN         6.111001e+13    0.0573

   transactiondate
0      2016-01-27
1      2016-03-30
2      2016-05-27
3      2016-06-07
4      2016-08-08

[5 rows x 60 columns]
```

Figure 6-11. *Merging few-to-many datasets*

Summary

This chapter explains the steps to clean and prepare data for analysis. It discusses how to rename the variable and adjust the reports' formatting. It also explains how to create new variables and how to rearrange them in datasets. It explains the IF and IF-ELIF statements. Moreover, it describes how to drop a row and a column, subset a dataset, and append and merge multiple datasets. Now, we are done with the most common data analysis tasks, and the next chapter will introduce predicting the future using regression.

CHAPTER 7

Regression

The linear regression task fits a linear model to predict a single continuous dependent variable from one or more continuous or categorical predictor variables. This task produces statistics and graphs for interpreting the results.

Simple Linear Regression

In simple linear regression, there is only one independent variable in the model. The line equation will be as follows:

$$Y = a + b\,X$$

where Y is the dependent variable that we want to predict its value, a is the intercept, b is the coefficient, and X is the independent variable. The error is ignored.

The null hypothesis is that there is no linear relationship between the variables. In other words, the value of b is zero. The alternative hypothesis is that there is a linear relationship, and b is not equal to zero.

$$H_0: b = 0$$

$$H_a: b \wedge= 0$$

Let us see if there is a linear relationship between the prices of oil per barrel and gold. The dataset is in the Datasets Folder with the name: sp_oil_gold.xlsx.

© Engy Fouda 2024
E. Fouda, *Learn Data Science Using Python*, https://doi.org/10.1007/979-8-8688-0935-4_7

There are several libraries that you can use in Python to perform linear regression. Two of the popular libraries are statsmodels and scikit-learn. The choice between using Ordinary Least Squares (OLS) from the statsmodels library and linear regression from scikit-learn (sklearn) depends on your specific needs and the context of your analysis. Both libraries offer linear regression functionalities, but they serve somewhat different purposes and have different features.

If you are primarily interested in statistical analysis, hypothesis testing, and detailed diagnostics, and you don't need the model for broader machine learning tasks, statsmodels might be a better fit.

If you are more focused on predictive modeling, integration with other machine learning tools, and simplicity, scikit-learn is a solid choice.

In some cases, analysts use both libraries: statsmodels for initial exploration and hypothesis testing and then scikit-learn for building a predictive model that can be easily integrated into a broader machine learning workflow. In our code, we shall include both scikit-learn and statsmodels, followed by diagnostic plots using seaborn to use the advantages of each one.

In the following example, we shall use the scikit-learn library for the model fitting and the fit diagnostics diagrams.

The steps to do the simple linear regression with its fit diagnostics are as follows:

1. Import the required libraries, load the dataset, and explore it.

2. Create the model.

3. Test the model and compute the predictions and accuracy score.

4. Plot five fit diagnostics for the oil price per barrel in
    the presidential election years:

    a. Predicted vs. Actual Values

    b. Residuals vs. Fitted Values plot

    c. R-Student vs. Predicted

    d. R-Student vs. Leverage

    e. Cook's D vs. Observation

The whole code is in Listing 7-1 in Chapter 7.ipynb in the Example
Code Folder. However, in the Jupyter Notebook, the code will be divided
in cells matching the above steps. Now, we shall explain each cell to
code each of the steps mentioned above with the explanation of the fit
diagnostics plots.

1. Import the Libraries, Load the Dataset, and Explore

In this step, we import the required libraries, load the dataset, and explore
it. To explore the data more, we plot the multiple pairwise bivariate
distributions of the variables.

Listing 7-1-1. Import the required libraries, load the dataset, and
explore it

```
import numpy as np
import pandas as pd
from sklearn.linear_model import LinearRegression
import matplotlib.pyplot as plt
import seaborn as sns
# importing r2_score module
from sklearn.metrics import r2_score
```

157

```
from sklearn.metrics import mean_squared_error
from sklearn.preprocessing import StandardScaler
import statsmodels.api as sm

df = pd.read_excel('../Datasets/sp_oil_gold.xlsx')

print(df.head())
df=df[['oil price','gold_prices']]
print(df.head())

# plot numerical data as pairs
plt.figure(0)
sns.pairplot(df);
```

In Listing 7-1-1, we start by importing nine libraries. Pandas, Numpy, and sklearn.preprocessing are for crunching the numbers and matrices. Matplotlib and Seaborn are for visualization. LinearRegression is a machine learning model. The sklearn.metrics libraries are for model evaluation. The statsmodels.api is to perform OLS and access influence statistics.

In Python, you can achieve the same task in multiple ways. Here, we loaded the data as a DataFrame using the pd.read_excel() function because the dataset is an Excel file. However, you could have loaded it as two Numpy arrays, as follows, because the dataset is small.

```
oil_prices = np.array([1.63, 1.45, 1.32, 1.82, 11.6, 35.52,
28.2, 14.24, 18.44, 20.29, 27.6, 36.05, 94.1, 109.45, 40.76])
gold_prices = np.array([35.27, 35.1, 39.31, 58.42, 124.74, 615,
361, 437, 343.82, 387.81, 279.11, 409.72, 871.96, 1668.98,
1250.74])
```

To verify that the data has been loaded corrected, we print the header and the first five rows of the dataset using df.head() function. We keep only the oil price and the gold_prices variables because this is a simple linear regression that needs only one independent variable, and we will drop the rest of the columns. We do that by statement:

```
df=df[['oil price','gold_prices']]
```

To explore the data, we print the pairwise bivariate distributions of the variables using sns.pairplot(df) function. To avoid the figures overlaying over each other, we enumerate them using plt.figure() function.

	Year	Stock Market Returns	returns	Republican won	oil price	gold_prices
0	1960	0.005	0.5	0	1.63	35.27
1	1964	0.165	16.5	0	1.45	35.10
2	1968	0.111	11.1	1	1.32	39.31
3	1972	0.190	19.0	1	1.82	58.42
4	1976	0.238	23.8	0	11.60	124.74

	oil price	gold_prices
0	1.63	35.27
1	1.45	35.10
2	1.32	39.31
3	1.82	58.42
4	11.60	124.74

```
<Figure size 432x288 with 0 Axes>
```

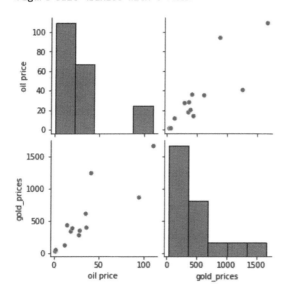

Figure 7-1. *Output of Listing 7-1-1*

Figure 7-1 shows the output of Listing 7-1-1. The figure shows at first the first five rows from all the columns in the dataset. Then, it shows the histogram of the oil price and the scatterplot of the oil price on the y axis with the gold_price on the x axis. In the second row of the figure distribution, the scatterplot is repeated again but with the gold_price on the y axis and the oil price on the x axis, and the second row and column is the gold_price histogram. In both scatterplots, there are three clear outliers, which we will discuss the reason causing this phenomenon.

2. Create the Model

In this step, we specify the independent and target variables. Because the dataset is small, we shall skip the training and testing splitting step. We shall use the whole dataset for training and testing. Then, we create the model and fit the data to it. From the parameter estimates, we shall write the line equation. Plot the fitted line visually overlaid over the variables' scatterplot.

Listing 7-1-2. Specify the independent and target variables

```
x=df['gold_prices'].values.reshape(-1, 1)
y=df['oil price'].values

# creating an object of LinearRegression class
LR = LinearRegression()
# fitting the data
LR.fit(x,y)

#parameter estimates
print('intercept:', LR.intercept_)
print('slope:', LR.coef_)
print('equation of oil prices in the presidential election
years=',LR.intercept_,"+",LR.coef_[0],"*gold_prices")
```

```
#Draw the fitted line
plt.figure(2)
plt.scatter(x, y)
plt.xlabel('Gold Prices')
plt.ylabel('Oil Price Per Barrel')
plt.plot(x,LR.intercept_+LR.coef_*x)
```

In Listing 7-1-2, we start with changing the independent variable, x, from DataFrame to a Numpy by using the .values property. As we mentioned before, there are multiple ways to do the same task in Python. We can use the method DataFrame.to_numpy() instead. The output of this property is a one-dimensional array.

However, the model fitting function requires a two-dimensional array. So, we use the reshape(-1,1) method to do so. The syntax of the reshape function is .reshape(number of rows, number of columns). To include all the rows without specifying a certain value, we use -1. Since this is a simple linear regression which means that we have one independent variable, we have 1 as the number of columns.

For the target variable, y, we again change its type from DataFrame to Numpy by using the .values property.

Now, our variables are ready to be fitted into the model. We first create an object of LinearRegression class using LR = LinearRegression(); then, we fit the variables using LR.fit(x,y).

From our model properties, we can get the parameter estimates and write the equation. We use them in plotting the line over the scatterplot of the two variables.

```
intercept: 1.9883089098584428
slope: [0.05964825]
equation of oil prices in the presedential election years= 1.9883089098584428 + 0.05964824505883558 *gold_prices
```

```
[<matplotlib.lines.Line2D at 0x285bc3667f0>]
```

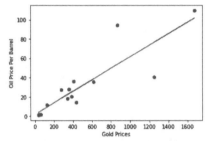

Figure 7-2. *Output of Listing 7-1-2*

Figure 7-2 shows that there is a linear relationship between oil and gold prices. The slope of the line = 0.05965. From the parameter estimate, the equation of the line is:

oil_price = 1.98831 + 0.05965 gold_prices

The plot consists of a scatterplot of data with the regression line overlaid over the scatterplot.

3. Test Model and Compute Predictions and Accuracy Score

Finally, we evaluate the model using R-squared, mean squared error, and root mean squared error.

Listing 7-1-3. Test the model, and compute the predictions and accuracy score.

```
#Listing 7-1-3
#Test Model:
y_prediction =  LR.predict(x)
print('y_prediction=',y_prediction)
```

```
# predicting the accuracy score
score=r2_score(y,y_prediction)
print('r2 socre = ',score)
print('mean_sqrd_error =',mean_squared_error(y,y_prediction))
print('root_mean_squared error =',np.sqrt(mean_squared_
error(y,y_prediction)))
```

In Listing 7-1-3, we computed the y_prediction values to calculate the accuracy of our model, which was implemented by the r2_score. It is a function inside sklearn.metrics module, where the value of r2_score varies between 0 and 100%.

Also, we compute the mean squared error (MSE), the estimator that measures the average of the squares of errors.

```
y_prediction= [   4.09210251    4.08196231    4.33308142    5.47295939    9.428831
   38.67197962   23.52132538   28.054592      22.49656853   25.12049483
   18.63673059   26.42738788   53.99919267 101.54003695   76.59275493]
r2 socre =   0.766030417790836
mean_sqrd_error = 228.6764484200374
root_mean_squared error = 15.122051726536231
```

Figure 7-3. *Output of Listing 7-1-3*

Figure 7-3 shows the output of Listing 7-1-3.

4. Fit Diagnostics

In Listing 7-1-4, we create diagnostic plots to assess the model's performance. In this code:

> R-Student vs. Predicted: This plot helps identify influential observations that may be driving the residuals.

> R-Student vs. Leverage: This plot helps identify observations with high leverage, which can disproportionately affect the model.

Cook's D vs. Observation: This plot helps identify
influential observations that may have a large
impact on the model parameters.

This part of the code uses statsmodels to fit an OLS model, and then,
it extracts influence statistics such as R-Student, leverage, and Cook's
D. The final part of the code creates diagnostic plots for R-Student vs.
Predicted, R-Student vs. Leverage, and Cook's D vs. Observation using
seaborn.

Listing 7-1-4. Fit diagnostics plots

```
#Listing 7-1-4

plt.figure(3)

# Plot diagnostic plots
plt.figure(figsize=(12, 5))
residuals = y - y_prediction

# Scatter plot of predicted vs. actual values
plt.subplot(1, 2, 1)
sns.scatterplot(x=y, y=y_prediction)
plt.title('Predicted vs. Actual Values')

# Residuals vs. Fitted Values plot
plt.subplot(1, 2, 2)
sns.scatterplot(x=LR.predict(x), y=residuals)
plt.axhline(y=0, color='r', linestyle='--')
plt.title('Residuals vs. Fitted Values')

plt.tight_layout()
plt.show()                   # Finalize and render the figure

# Use statsmodels to perform OLS and access influence
statistics
```

```python
X_ols = sm.add_constant(x)
ols_model = sm.OLS(y, X_ols).fit()

# Access influence statistics
influence = ols_model.get_influence()
r_student = influence.resid_studentized_internal
leverage = influence.hat_matrix_diag
cook_d = influence.cooks_distance

# Remove missing values
mask = ~np.isnan(r_student) & ~np.isnan(leverage)
r_student = r_student[mask]
leverage = leverage[mask]

# Plot R-Student vs Predicted
plt.figure(figsize=(12, 4))
plt.subplot(1, 3, 1)
sns.scatterplot(x=LR.predict(x), y=r_student)
plt.axhline(y=0, color='r', linestyle='--')
plt.title('R-Student vs Predicted')

# Plot R-Student vs Leverage
plt.subplot(1, 3, 2)
sns.scatterplot(x=leverage, y=r_student)
plt.axhline(y=0, color='r', linestyle='--')
plt.title('R-Student vs Leverage')

# Plot Cook's D vs Observation
plt.subplot(1, 3, 3)
#print(len(df.index),len(cook_d))
sns.scatterplot(x=df.index, y=cook_d[0])
plt.xlim(0, 16)  # Set y-axis limits to have values above and
below zero
# Set x-axis ticks to increment by 1
```

```
plt.xticks(np.arange(min(df.index), 15, 1))
plt.axhline(y=0, color='r', linestyle='--')
plt.title("Cook's D vs Observation")

plt.tight_layout()
plt.show()
```

These plots are based on statistical diagnostics and can provide insights into the influence of individual observations on the model. Adjust the code as needed for your specific analysis and requirements. Let us look at the output and explain them.

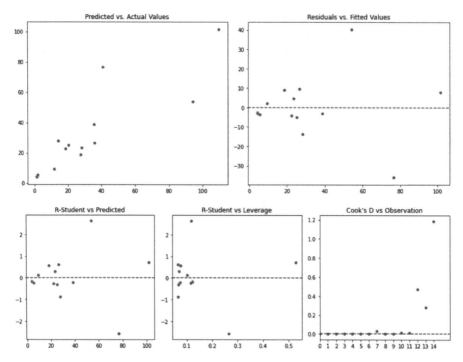

Figure 7-4. *Output of the fit diagnostics plots*

Figure 7-4 shows the fit diagnostics for `oil_price`. The Dependent Variable (task) vs. the Predicted Value plot visualizes variability in the prediction. In this plot, the dots are random, and there is not a pattern that indicates a constant variance of the error. The outliers are clear in this plot.

Residual vs. Fitted shows a random pattern of dots above and below the 0 line, which indicates an adequate model. Again, the three outliers are clear here.

The R-Student vs. Predicted Value plot shows a couple of outside the ±2 limits. A third dot lies in between them but is far away from the rest of the dots.

The R-Student vs. Leverage plot shows the outliers which have leverage on the calculation of the regression coefficients.

The Cook's D plot is designed to identify outliers. Here, the three points are so clear with their values. They are at Rows 12, 13, and 14. We shall explore these outliers and the reason behind them in the next section.

These diagnostic plots help assess the assumptions and performance of the linear regression model. They include checking for linearity, homoscedasticity (constant variance of residuals), and influential observations. The plots can provide insights into the model's behavior and identify potential issues that might require further investigation or model refinement.

Multiple Linear Regression

In multiple linear regression, there is more than one independent variable in the model. The line equation will be as follows:

$$Y = a + b_1 X_1 + b_2 X_2 + \ldots + b_i X_i$$

where Y is the dependent variable that we want to predict its value, a is the intercept, b_i are the coefficients, and X_i are the independent variables. The error is ignored.

The null hypothesis is that there is no linear relationship between the dependent variable and any of the independent variables. In other words, the values of b_i are zeros. The alternative hypothesis is that there is a linear relationship and any b_i does not equal zero.

$$H_0: b_i = 0$$

$$H_a: b_i \wedge= 0$$

Let us see if there is a linear relationship between the stock performance index (S and P index) and the prices of oil per barrel and gold in the presidential election years. The dataset is in the Datasets Folder with the name: sp_oil_gold.xlsx. The code is in Listing 7-2 in Chapter 7.ipynb in the Example Code Folder.

Listing 7-2 displays three vertical bar charts showing the prices of oil, gold, and stocks over the years (from 1960 to 2016), each in a dedicated graph. The years are the presidential election years.

Listing 7-2. Exploring the stock, oil, and gold prices

```python
import numpy as np
import pandas as pd
from sklearn.linear_model import LinearRegression
from matplotlib import pyplot as plt

df = pd.read_excel('../Datasets/sp_oil_gold.xlsx')

oil_price = df['oil price']
gold_prices = df['gold_prices']
Stock_Market_Returns=df['Stock Market Returns']
years=df['Year']

# Figure Size
plt.figure(0)
# set width of bar
barWidth = 0.25
```

```python
fig = plt.subplots(figsize =(20, 8))
plt.xticks(years)
plt.yticks(Stock_Market_Returns)
plt.axhline(y=0, color='b', linestyle='-')
plt.title('Stock_Market_Returns')
# Bar Plot
plt.bar(years,Stock_Market_Returns)

# Show Plot
plt.show()

plt.figure(1)
 # set width of bar
barWidth = 0.25
fig = plt.subplots(figsize =(20, 8))
plt.xticks(years)
plt.yticks(oil_price)
plt.title('oil_price')
plt.axhline(y=0, color='b', linestyle='-')

# Bar Plot
plt.bar(years,oil_price)

# Show Plot
plt.show()

plt.figure(2)
 # set width of bar
barWidth = 0.25
fig = plt.subplots(figsize =(20, 8))
plt.xticks(years)
plt.yticks(gold_prices)
plt.title('gold_prices')
plt.axhline(y=0, color='b', linestyle='-')
```

```
# Bar Plot
plt.bar(years,gold_prices)
```

```
# Show Plot
plt.show()
```

```
x=df[['oil price','gold_prices']].values
y=df['Stock Market Returns'].values
```

```
# creating an object of LinearRegression class
LR = LinearRegression()
# fitting the training data
LR.fit(x,y)
```

```
print('intercept:', LR.intercept_)
print('slope:', LR.coef_)
print('equation of Stock market returns=',LR.intercept_,"+",LR.
coef_[0],'* oil price +',LR.coef_[1],"*gold_prices")
```

The output will be as in Figures 7-5, 7-6, 7-7 and 7-8. The 2008 financial crisis is clear in all the graphs. This crisis is the worst economic disaster in the United States since the Great Depression of 1929. The oil and gold prices spiked in 2012.

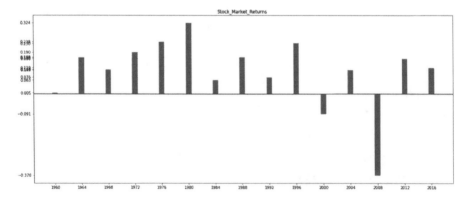

Figure 7-5. *Exploring S and P index*

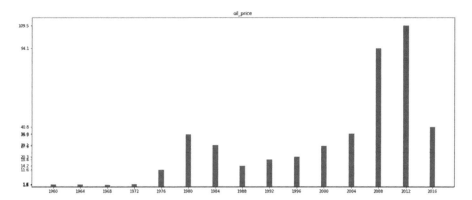

Figure 7-6. *Exploring oil price*

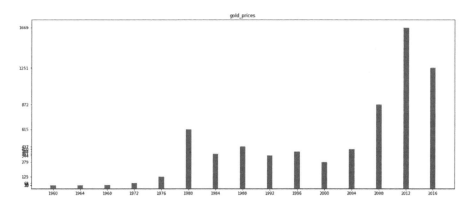

Figure 7-7. *Exploring gold price*

Despite these three values acting as outliers, we cannot exclude them because they have a direct impact on how public might vote. Next, as we did in the previous section, we create the model and fit the data to it. Finally, we computed the parameter estimates, the slope, and the intercept.

From the parameter estimate, the equation of the line is:

stock_market_returns = 0.13010 - 0.00671 oil_price + 0.00036367 gold_prices

```
intercept: 0.13010280357814358
slope: [-0.0067133  0.00036367]
equation of Stock market returns= 0.13010280357814358 + -0.006713300291729812 * oil price + 0.00036367435673556966 *gold_prices
```

Figure 7-8. *Output*

Figure 7-8 shows the output of the parameter estimates and the equation.

You can repeat the same steps of the last section to plot fit diagnostics. However, we find that this is enough for demonstrating the concept of the multiple linear regression.

Logistic Regression

Logistic regression is similar to linear regression except that the dependent variable is of binary values, not continuous. The model equation checks the probability of one of the discrete binary values of the dependent variable occurring.

The binary logistic regression model equation is as follows:

$$p=e^{\wedge\ (b0+b1X1+b2X2+...)} / (1+e^{\wedge\ (b0+b1X1+b2X2+...)})$$

Again, the null hypothesis is that there is no relationship between the variables, so all the coefficients are zeros. The alternate hypothesis is that there is a relationship between the dependent variable and the independent variables.

For this example, we use the Titanic dataset to see if there is a relationship between the passenger's survival, age, and passenger's class as independent variables. The dataset is in the "Dataset" folder and has the name titanic.csv, or you can download the CSV file from OpenDataSoft. com at the following URL: https://www.kaggle.com/datasets/ vinicius150987/titanic3/data.

The code is in Listing 7-3 in Chapter 7.ipynb in the Example Code Folder.

Listing 7-3-1. Logistic regression

```
#Listing 7-3-1
import pandas as pd
from sklearn.linear_model import LogisticRegression

df = pd.read_csv('../Datasets/titanic.csv')
print(df.head())

df['male'] = df['Gender'] == 'male'

x = df[['Pclass', 'male', 'Age', 'Siblings/Spouses', 'Parents/
Children', 'Fare']].values
y = df['Survived'].values
print('x[0:5]=',x[0:5])
print('y[0:5]=',y[0:5])
```

In Listing 7-3-1, we start by importing the required libraries. Then, we load the data; because the dataset is in CSV format, we use pd.read_csv() function. As in the previous sections, we check if the dataset has been loaded correctly in the DataFrame, df, by printing the first five rows. The output is shown in Figure 7-9.

Let us look at the output of this code before proceeding explaining the code.

```
   Survived  Pclass  Gender   Age  Siblings/Spouses  Parents/Children      Fare
0         0       3    male  22.0                 1                 0    7.2500
1         1       1  female  38.0                 1                 0   71.2833
2         1       3  female  26.0                 0                 0    7.9250
3         1       1  female  35.0                 1                 0   53.1000
4         0       3    male  35.0                 0                 0    8.0500
x[0:5]= [[3 True 22.0 1 0 7.25]
 [1 False 38.0 1 0 71.2833]
 [3 False 26.0 0 0 7.925]
 [1 False 35.0 1 0 53.1]
 [3 True 35.0 0 0 8.05]]
y[0:5]= [0 1 1 1 0]
```

Figure 7-9. *Output of Listing 7-3-1*

All the columns of this dataset are numeric except the Gender, which is string. We cannot use a string variable in regression. There are multiple ways to encode the string variables to numeric, for example, the ordinal and one-hot encodings. However, in our case, we do not need these encodings because the gender has two values only, male and female. So, it is easier to add a mask on one of the values, which means if the mask is true, the value will be 1; otherwise, it will be 0.

We create a Pandas Series that will be a series of True and False values (True if the passenger is male and False if the passenger is female).

```
df['Gender']=='male'
```

Now, we want to create a column with this result. To create a new column, we use the same square bracket syntax df['male'] and then assign this new value to it.

```
df['male'] = df['Gender'] == 'male'
print(df.head())
```

Next, we set the independent variables in DataFrame x and the target variable Series y for the passengers who survived Titanic. In the code, we print the first five rows of both x and y.

Listing 7-3-2. The model

```
#Listing7-3-2
model = LogisticRegression()
model.fit(x, y)
print('model.coef_=',model.coef_)
print('model.intercept_=',model.intercept_)
```

In Listing 7-3-2, we create an instance of the LogisticRegression and fit the x and y to it. Then, we print the parameter estimates of the model, as in Figure 7-10.

```
model = LogisticRegression()
model.fit(x, y)
print('model.coef_=',model.coef_)
print('model.intercept_=',model.intercept_)
```

```
model.coef_= [[-1.13645583 -2.6440994  -0.04237384 -0.38746621 -0.09619821  0.00297187]]
model.intercept_= [5.08856862]
```

Figure 7-10. *Output of Listing 7-3-2*

Figure 7-10 shows the parameter estimates, and you can substitute the values in the equation of the logistic regression.

Listing 7-3-3. Test the model and compute the accuracy

```
#Listing 7-3-3
print('The predictions of the first 5 rows=',model.predict(x[:5]))
# [0 1 1 1 0]
print('The actual y values of the first 5 rows=',y[:5])
# [0 1 1 1 0]
y_pred = model.predict(x)
print('accuracy=',(y == y_pred).sum()/y.shape[0])
print('model score=',model.score(x,y))
```

In Listing 7-3-3, we simply compute the predictions and the difference between the model predictions and the actual values. This computation is the model accuracy. There is function of the logistic regression called model.score() that makes this difference automatically without the need to compute the predictions manually.

```
The predictions of the first 5 rows= [0 1 1 1 0]
The actual y values of the first 5 rows= [0 1 1 1 0]
accuracy= 0.8049605411499436
model score= 0.8049605411499436
```

Figure 7-11. *Output of Listing 7-3-3*

Figure 7-11 shows the output of Listing 7-3-3. In the figure, we compare the manual computing of the accuracy and the output of the `model.score()` function. As shown, there is no difference between them, and this model's accuracy is about 81%.

Summary

This chapter explains single, multiple linear, and the logistic regressions. It explains in detail the diagnostic matrix plots and the meaning of each of them. Moreover, it shows how to easily spot the outliers in several plots of the diagnostic plots.

Index

A, B

Anaconda installation, 23–24
Arithmetic and assignment
    operators, 106–107

C

Comment statement, 105–106

D, E, F, G

Data analysis, 1
  age, 9
  categories, 14, 15
  economy, 15
    exit polls, 17
    historical data, 17, 18
    income sources, 15
    median income data, 16
    political trajectory, 19
    unemployment rate, 16
  genders, 6, 7
  historical voter turnout, 10, 11
  issues, 14, 15
  population distribution, 3–6
  presidential elections, 3, 13
  process, 2, 3
  process modeling

democratic party, 21
    2016 prediction results, 20, 21
    2020 prediction results, 21, 22
    statistical tests, 19
  question/hypothesis, 2
  racial demographics, 8
  raw data, 2
  statistical tests, 1
  trend measurement/pattern
    analysis, 2
  winning percentages, 11–13
Data cleaning, 2, 5, 105, 125
Data preparation/processing
    techniques, 123
  appending datasets, 148–151
  date/currency
    formatting, 127–130
  dictionary, 123
  DROP method, 143–145
  dropping rows, 142, 143
  IF-ELIF statements, 138–142
  IF statement, 137–139
  merge operation, 151–153
  rearranging variables, 134, 135
  rename method, 124–127
  subset output/frequency
    table, 148–150
  variable creation, 131–134

E. Fouda, *Learn Data Science Using Python*, https://doi.org/10.1007/979-8-8688-0935-4

V, W, X, Y, Z

Printed in the United States
by Baker & Taylor Publisher Services